EUROPE'S GIANT ACCELERATOR

The Story of the CERN 400 GeV
Proton Synchrotron

The culmination of eight years negotiation and five years design and constructional effort. Dr. J. B. Adams, in June 1976, surrounded by members of his international team of engineers and physicists. They have the satisfaction of seeing the machine they have built achieve its design energy.

EUROPE'S GIANT ACCELERATOR

The Story of the CERN 400 GeV
Proton Synchrotron

*Maurice Goldsmith
and
Edwin Shaw*

TAYLOR & FRANCIS LTD . LONDON 1977

First published 1977 by Taylor & Francis Ltd
10–14 Macklin Street, London WC2B 5NF

© 1977 M. Goldsmith and E. N. Shaw

All rights reserved. No part of this publication may be reproduced, stored in a retrieval system, or transmitted, in any form or by any means, electronic, mechanical, photocopying, recording or otherwise, without the prior permission of the copyright owners

ISBN 0 85066 121 8

Printed and bound in Great Britain by
Taylor & Francis (Printers) Ltd
Rankine Road, Basingstoke, Hampshire
Design and production in association with Book Production Consultants,
7 Brooklands Avenue, Cambridge

Contents

Foreword — vii

Introduction — ix

  1/The Long Search — 1
  2/The Practical Dreamers — 21
  3/Site for Agreement — 43
  4/Enter the Builders — 63
  5/Down to Earth — 91
  6/Guiding and Focusing — 115
  7/Power and Water — 135
  8/Ins and Outs — 153
  9/Acceleration — 165
10/Machine Management — 181
11/Start-Up — 205
12/A Tool to Use — 221
13/Epilogue: the Long Search Continues — 237

Appendixes
 I/Adams on Accelerators — 239
 II/Contracts listed by Country in Order of Value
   (above 100 000 SF) — 253

Acknowledgments — 256

Index of Names — 257

Subject Index — 259

Professor Edoardo Amaldi of the University of Rome, presides over the CERN Council meeting held on 19 February 1971 at which the decision was taken to build in Europe the world's biggest tool—the SPS. Professor Amaldi was prominent in the early discussions that led to the establishment of the provisional Conseil Européen pour la Recherche Nucléaire in 1952 of which he became Secretary-General. Throughout the subsequent history of CERN, his guidance and counsel have been major factors in the success of the Organization. He became the first chairman of the European Committee for Future Accelerators in 1963, presenting the report that led, eight years later to the agreement to build the SPS. Member of the Scientific Policy Committee and Italian delegate to Council over many years, he is still seen regularly at CERN working in the laboratory with a team of experimental physicists.

Foreword

The completion of the building of the Super Proton Synchrotron of CERN, its operation and the start of its utilization for scientific work before the end of 1976, are the culmination of a long process which began in 1960 at the level of preliminary discussions, evolved around 1963 into a programming phase and in 1971 entered the construction period.

The remarkable success of this endeavour is mainly due to the excellence of the work of the staff of CERN who were chosen, organized and led to their final goal through six years of intensive day-to-day effort by Dr. J. B. Adams.

The SPS incorporates many advanced features of engineering design in both the system concept and the realization of its separate components. It is the first big machine, for example, to be entirely controlled through computers; the accelerating system is quite novel; the ejection systems have required the development of new techniques; the civil engineering has attained new levels of precision; even the power supplies that couple this European machine to the European electricity grid network are highly original.

The scientific programme of research is now under way. This too is the result of long preparatory discussions, computation, design and meticulous fabrication, carried out by physicists all over Europe and beyond. The energy region being covered by the SPS to begin with is already being explored by the Enrico Fermi National Accelerator Laboratory in the USA, but so wide is the field to be studied, there is plenty of room for two laboratories as big as FNAL and CERN to search and discover. Even so, neutrino and muon physics, both large areas of fundamental importance, should be better done at the CERN laboratory because of the better quality of the beams and the size and diversity of the detectors.

Returning though to the SPS, the construction of the machine was an enterprise whose exceptional features were not limited to the technological innovations, or even to the scale of its conception and financing. This was an enterprise that was also quite exceptional in the collaboration that was manifest between countries, between physicists, diplomats, engineers and administrators.

Foreword

The story of this enterprise starting from the early discussions on the future of CERN through to the beginning of the experimental programme, is presented in this book, written by M. Goldsmith and E. N. Shaw, who was for nine years responsible for the Public Information Office of CERN and particularly well placed to follow the progress of the project from day to day. The authors had at their disposal all the necessary documentation and they succeed, I think, in transmitting to the reader the importance of this great enterprise.

This importance stems not only from the scientific and technical aspects but also from the highly successful collaboration of 12 Western European countries. These nations have shown again their ability to work together, creating on an even bigger scale, a remarkable instrument of research, around which scientists from all over the world will meet and cooperate for the next 20 years and more. Having been involved personally in this enterprise during its initial phases, I am very happy now to see its successful conclusion and I am full of admiration for the magnificent work of Dr. Adams and his staff. I am grateful to them for this machine that accelerates protons and distributes them to the experimentalists, and which they have designed and constructed in a way that exceeds our most optimistic expectations.

All of us hope that this extraordinary achievement may also help to convince the leaders of our society that advanced problems can be solved by a common effort. The success of European collaboration at CERN is not an accident, nor is it due solely to the fact that CERN was created to do pure science. It is due also to the structure of the organization, which encourages the commitment of everyone to the pursuit of a common goal, but has room for understanding when an individual partner has special problems. In CERN, we have a model in Europe from which all countries can learn.

Edoardo Amaldi
January 1977

Introduction

When twelve European nations agree together to construct the biggest single machine in the world and this machine is designed and built within the projected time-scale and the budget estimates, attaining its full design specifications even before the date foreseen, this is an event of note and one in which Europeans can take pride. A major achievement of our generation is the bringing into operation of the CERN 400 GeV Super Proton Synchrotron or SPS. It continues a series of successes that have marked the progress of the European Organization for Nuclear Research since its birth in the early 1950s.

The name of the machine suggests an esoteric venture, incomprehensible to the lay person, of little relevance to the major pre-occupations of our time. Built to provide the high energy physicists of Europe with the tool necessary for them to continue their studies of the fundamental structure of matter, it will only incidentally affect the price of goods sold in our shops or the material comforts of those who have sponsored the project. Nothing, however, is more fundamental to our existence than our understanding of the elementary constituents of our Universe and the laws which govern their behaviour.

This book tells a part of the story of the evolution of the project to build in Western Europe the machine that is the largest tool constructed by man. It was not an easy project to formulate and only through the determination of the scientists, and their readiness to make sacrifices to arrive at a common goal, was it able to surmount the difficulties inherent in its conception. Equally important was the tenacity of the government representatives and their spirit of cooperation through the long period of discussion that led to the final agreement. Science and government united to create a laboratory that was beyond the capacity of any single nation.

In telling this story we have tried to understand, and then explain, the essential technical aspects of the machine around which the laboratory is centred, in a way that is accessible to readers unfamiliar with the jargon of high energy physics, or the techniques of the machine builders. The experts may, at times, be offended by our simplifications, but we take comfort from the fact that few scientists even

Introduction

are cognizant with all the disciplines that are involved, and specialists are, for the most part, specialists in narrow domains only.

In its essence, the story of the SPS is the story of an engineering project. Engineering is the realization of a defined objective at the minimum cost, compatible with social needs. Building the SPS required a clear understanding of the use to which the finished machine would be put. The SPS is not an exercise in applied ingenuity, but a machine with an explicit function to perform—to provide, reliably and economically, beams of high energy particles of pre-determined characteristics for physics experiments. The builders needed to have an intimate knowledge of the experimentalists' requirements and to maintain a continuous dialogue with those who would be exploiting its potentialities. Ingenuity and inventiveness are to be found in abundance in the machine's design, but they are there to satisfy the exigencies of the programme. From the outset, the time-scale and budget profile were clearly defined and considered as absolute boundary conditions. Satisfying these conditions required inspired leadership and efficient management, which could draw out the best from the international team that was assembled, exploiting the genius of the individuals and making allowance for their weaknesses. Even this would not have been enough if the result had been a complex of ugly constructions that were a danger or nuisance to the inhabitants of the region, or those employed on the site. Such was not the case; the safety of the local population and workers and the protection of the environment, formed an integral part of the laboratory design.

Our story of the SPS ends with the drama of the last days leading up to the achievement of full design energy. We must, however, also place on record that since that date in June, 1976, full intensity was reached at maximum energy in October, and the experimental programme got under way the following month, so fulfilling, in an exemplary manner, the promises that had been made six years before.

If this book reads as a tribute to Dr. J. B. Adams and his team, this is entirely due to the success of their collective endeavour; we have simply tried to put down the facts as we saw them.

The views and comments expressed in this book are entirely our own, and in no way express official CERN opinion.

Maurice Goldsmith
Edwin Shaw
January 1977

1/The Long Search

Our desire to understand more of the environment in which we live, to probe ever deeper into Nature and to unveil its mysteries, is fundamental to our existence. It is a mainspring of our culture, at the very root of our evolution in the arts as well as the sciences. Although scientific development has brought about dramatic changes in our standards of living and our way of life, the fundamental advances have been born of an inner urge for understanding.

Research is an activity which may take many forms. The scientific approach of making concentrated attacks on closely defined problems has, over the second half of this millenium, met with spectacular success. Precise and reproducible information has been gathered, whose interpretation has led to the formulation of broad generalizations which have then been tested by rigorous experiment. Building on those theories that experiment has verified, and discarding those that proved to be inadequate, science has been able to make immense intellectual progress in our interpretation of what constitutes the Universe. At regular intervals, cherished beliefs, that once seemed immutable, have been superseded: physics has had to invent completely new ways of thinking. This willingness to re-think, to build on the past, but to reject its conclusions when these no longer stand the test of experiment, are the essential safeguards of science.

Physics is an art of the possible; experiments are done which can be done. Although the general direction in which a given line of research should be pursued is indicated by the intellectual pressures of the past, the techniques within our reach must dictate the methods employed. Sometimes a number of possible routes are open, and the choice between them may then be determined by considerations which reflect the scientists' personal interests; but among physicists concerned with the study of the smallest components of matter, the elementary particles, there is a world-wide consensus of opinion that the unique way open at the present time is to study the behaviour of these particles by the use of high energy accelerators.

Particle physics and high energy physics are terms which have become synonymous. In recent years, East and West Europe and the United States have taken it in

Europe's Giant Accelerator

turns to build accelerators with the highest energy, and so push back a little further this frontier of knowledge. China may well become a fourth participant in a field of research where international collaboration and competition exist in harmonious relationship. Faced with such a solid body of scientific opinion, the non-physicist has no option but to believe that, if we are to continue research into the fundamental constituents of Nature, there is no practical alternative to the construction of very large machines—machines which are the largest tools in the world.

Behind all physics research are two great acts of faith; we believe that nature is comprehensible, and we believe that nature plays fair. So far, both these assumptions seem justified. There may be areas of truth which cannot be comprehended, but, within its limited field, physics has met no suggestion to this effect, and Nature has always been found to 'play the game'; physics, as a result, is an exact science. This does not mean that all things at all times can be exactly predicted. Chance plays a role, and even a dominant role at the individual level, but global rules can be found which, if they are obeyed here and now, will be obeyed tomorrow somewhere else. The apparent universality of the laws of physics, which allows researchers in the laboratory to calculate with precision the behaviour of bodies

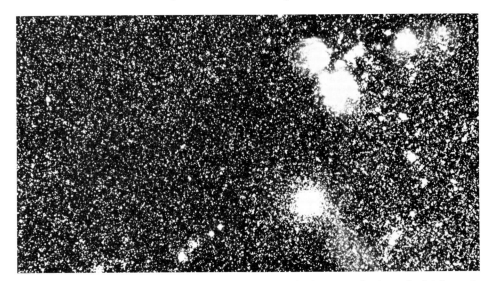

It is a remarkable and comforting fact that scientific laws can be found which apply here and now, yet apply equally rigorously to events which occurred in the distant past billions of light years away. Physics is the science of the universal laws of Nature. (Photo ESO)

1/The Long Search

millions of light-years away, is something to marvel at and be grateful for. If it were not so, we should be living in a Universe that was fundamentally chaotic, whereas we can be confident that at the basic level of our existence, order reigns.

Over this century, physics has come to recognize that matter is not an anonymous smear of infinitely continuous properties. It is built of individual constituents of astonishingly consistent characteristics. Nature works with collections of individuals, which are independent and interdependent at the same time. In consequence, the physicist is obliged to work with populations, inferring the essential qualities of the individual by the observation of mass movements, or by the statistical analysis of patterns of behaviour in which an unknown number of variables are involved. He must be aware also of his own limitations, conscious of the fact that the very act of observing may disturb the system he observes to the point that some of its fundamental features are obscured.

In identifying these fundamental features and the essential rules which govern behaviour, we start with neither a definition of the individual, nor a description of what features are significant. If, at each new level of knowledge, Nature re-invented new rules and new concepts, progress in physics would be very slow, but

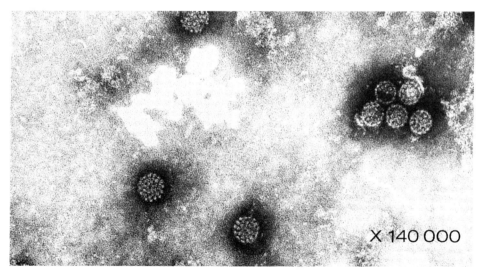

These living virus particles, like inanimate objects, are composed of ordered patterns of molecules in turn composed of ordered patterns of atoms. The atoms themselves can be broken down into a small number of components whose characteristics determine the physical structure of our Universe. (Photo D Chescoe)

Europe's Giant Accelerator

Nature works with populations of individuals. The individuals have a clear identity, but a degree of independence which makes it impossible to forecast their exact behaviour, even though exact laws can be found which statistically describe their interactions. Their identity may depend as much on the society in which they find themselves as the society depends on its component members. (Photo Camera Press)

whilst there are dramatic changes to be observed, there is an underlying consistency. Ideas and techniques that have worked in the past give clues to the future, even if they do not provide the complete answers.

From time to time, intellectual leaps are needed, one of the most important of which was the realization that, at the most elementary level, matter possessed certain qualities not in arbitrary, but in very strictly controlled, amounts (quanta) which could be quantified as simple numbers. Quantum mechanics was developed first to help us understand the nature of the atom, but it also gave us the vital lead in the study of elementary particles. It suggested that at this lower level, matter would possess qualities which, if we could recognize them, would be found to occur in standardized packets. The problem was to discover what these qualities were, and what constituted a packet. When this was understood, it should be possible to work out the rules under which the possessors of the packets interacted, the nature of their individuality, and their relationship with their environment.

The high energy physicist is continuing a pattern of research that goes back to Stone Age man. When he looked at the trees and the bushes and began classifying them into evergreens and deciduous, edible fruit-bearing and non-edible fruit-bearing, he was beginning a theoretical classification of a kind that has an important place in modern research. Although unaware of the mechanism he was using, Stone Age man, when looking at his rocks and trees, was analysing the amplitude (brightness) and frequency (colour) of the electromagnetic radiation (light) scattered by the objects of his attention when these were bombarded with elementary projectiles (light quanta emitted by the sun). When he was smashing together his flints, he was accelerating one mass to a high energy, projecting it against a target, and analysing the collision products.

Stone Age man in striking one object against another was beginning a line of research that has led to high energy physics, where new discoveries are made by bringing elementary particles into violent collision and observing the results.

Europe's Giant Accelerator

A powerful technique in research is to observe the behaviour of natural phenomena and to classify the results. The botanist looks for the similarities between growing objects and the physicist looks for similarities between the elementary particles. The patterns formed by the similar groups, indicate the qualities which have a distinguishing significance. (Photo Camera Press)

There is a limit to the concentration of energy that can be obtained by mechanical means. For research into the structure of matter it is not the total energy available which counts, but the energy that can be imparted to the individual constituents. Below the level of crystals, more subtle means have to be employed. (Photo Camera Press)

Europe's Giant Accelerator

CERN seen from above the Jura mountains, looking across Geneva to the Alps and Mont Blanc in the distance. Left of centre is Lac Léman, with the Rhône flowing out to the right. The triangular-shaped laboratory of the older CERN is a little left of centre, and the laboratory area of the SPS away to the left. (Photo Swissair)

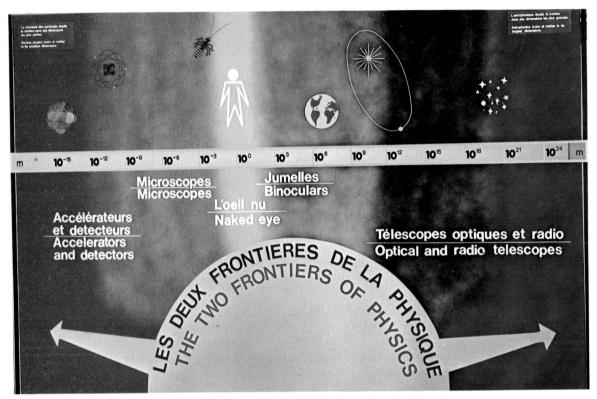

The limits to observation stretching from the very edge of the Universe to the smallest dimensions accessible to us. For distant objects, radio and optical telescopes gather information which is interpreted in terms of the composition and age of the stars, suggesting that all originated from some primordial explosion. Over the centre range, direct viewing down to sizes of the order of a millionth of a metre give us the illusion of a more exact analysis because of our familiarity with shapes and colours. Our knowledge of our environment comes, however, from a much wider range of experience than simply looking. To penetrate to the smallest dimension, accelerators are the only tools available.

Europe's Giant Accelerator

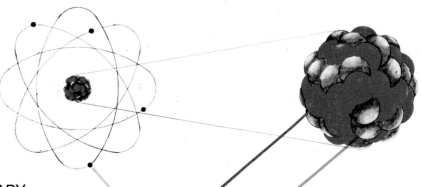

atom   nucleus

ELEMENTARY PARTICLES STUDIED AT CERN
{ ELECTRON e PROTON p NEUTRON n stable or long-lived and many others mostly unstable

CLASSIFICATION   these numerous elementary particles may be grouped in families:

family of the proton: p n $\Delta°\Sigma^+\Sigma°\Sigma^-\ \Xi^-\ \Xi°$ (6 unstable particles)
family of the electron: e μ (muon unstable)
family of the neutrino
and others...

An atom is made up of a tiny central nucleus, carrying almost all the mass, surrounded by a cloud of electrons whose electric charge exactly balances the charge carried by the protons in the nucleus. In the nucleus also are neutrons, particles similar to the protons, but with no electric charge. Outside the atom, two forces are felt: one from the electrical charges within the atom; the other from the gravitational attraction of the mass. Inside the nucleus, two other forces—the strong and the weak—are at work holding it together, and determining if and when a neutron changes into a proton. At the sub-nuclear level, there are many other particles with an ephemeral existence. Classifying the particles into families helps us to understand their essential qualities.

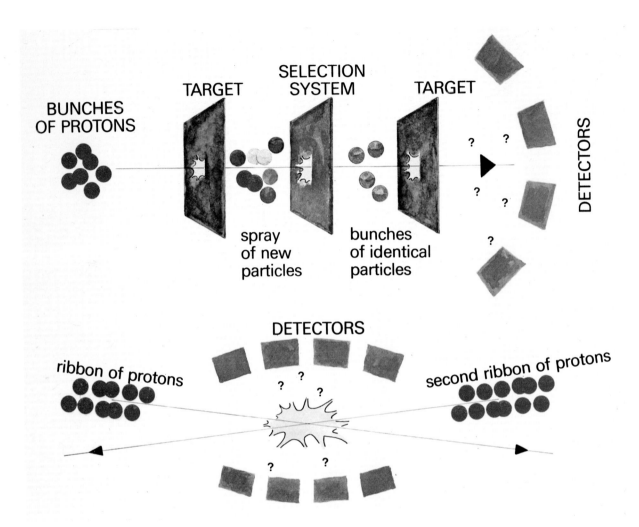

Looking at an object involves its bombardment with photons (light), and observing the characteristics of the scattered radiation. In the most elementary high energy physics experiments, bunches of protons accelerated to high energy are directed on to a target, and detectors record the results. It is more interesting to observe the reactions which take place when less familiar particles collide with the target in a second stage. In the first target, secondary particles are produced. These are analysed and selected, and a beam of them is directed on to a second target surrounded by detectors. In the ISR, however, we observe the reactions of protons on protons, but beams of secondary particles cannot be produced.

PRINCIPAL PARTICLE DETECTORS

– – –▶ – – –   particle trajectory (invisible)
• • • • • •   visible tracks

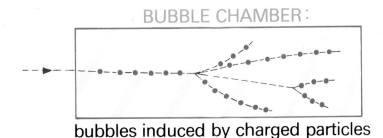

BUBBLE CHAMBER:
bubbles induced by charged particles

COUNTERS, SPARK CHAMBERS:
sparks induced by charged particles

The two principal particle detectors are the bubble chamber and the spark chamber. In the former, the working liquid is also the target. Lines of bubbles form along the tracks of charged particles passing through, whilst uncharged particles leave no trace. Their presence can be inferred when they transform into pairs of charged particles, or collide with a proton in the liquid; also, by analysing the reaction. If the chamber is surrounded with a magnet, the tracks are curved, the degree of curvature indicating the energy the particle is carrying.

Rows of spark chambers allow the charged products from a collision to be detected, and by interposing a magnetic field, or enclosing the array in a magnet, information can be obtained about the energy. Each event is unique and unreproducible, and only by observing a great number of events can the essential characteristics be deduced.

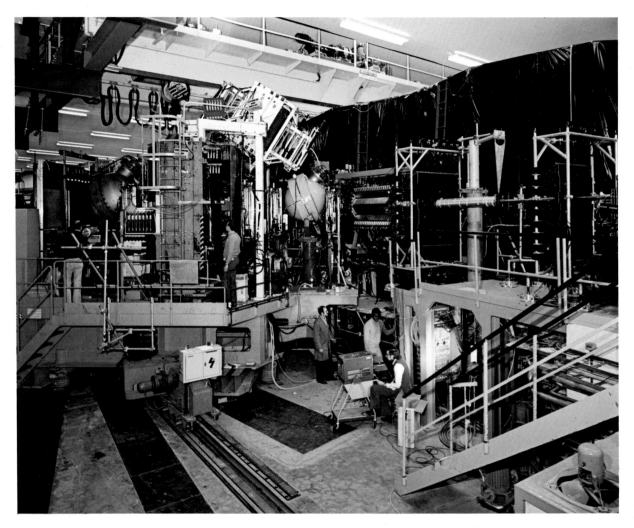

In a typical electronics experiment, many detectors will be grouped round the target to limit the number of explanations of the behaviour of the particles observed. Because the total information from any event is strictly limited, much of the equipment is designed to exclude the recording of unwanted reactions. Frequently, the rare event is the most revealing, and the problem is to detect one particular set of particles. Many of the detectors are there to signal when the main detectors should record, and a very large quantity of data has to be processed to sift out the interactions of significance.

It is convenient at times to take pictures of spark chamber events and record them on film. The two 'wings' in the photograph are spark chambers about two metres in length.

One of the largest bubble chambers in the world filled with liquid hydrogen is BEBC. The body of the chamber, being lowered into position, is a stainless steel domed cylinder 3·7 metres diameter, surrounded by two superconducting coils, cooled by liquid helium, that generate a high magnetic field inside the chamber. A massive iron shield surrounds the assembly to contain the field. From below, a piston working in a cylinder, controls the pressure in the chamber. Cameras mounted above the port-holes on the top record the events inside.

1/The Long Search

In spite of the refinement of the accelerating equipment, and the detectors which record the results, the experimental techniques of high-energy physics are, in essence, similar. One group of particles is projected against another, and the scattered products are observed. Only by repeating the same type of experiment over and over again, with as few variables as possible, and classifying and analysing the results, can the fundamental patterns be made to emerge.

Stone Age man was able to break up his masses of rock into individual pieces because the bonds holding them together were relatively feeble, so that the energy that he had to put in was relatively modest, and the object with which he was concerned was large. But as we probe deeper into the constitution of matter, so the energy put in must be concentrated more and more; it is not so much the total quantity, as the way in which it is applied. Increasing the applied force does not necessarily help. The sledgehammer on the proverbial nut extracts little more juice, as the greater part of the energy is not absorbed by the components of the nut, but goes into the anvil on which it is sitting.

Heat and chemical reactions, which impart energy at the molecular and atomic levels, concentrate the energy in a way that leads to the breakdown of molecules and the re-ordering of atoms. The outer structure of the atoms is changed but the inside remains untouched, unless temperatures of millions of degrees are produced. (Photo Camera Press)

'Seeing' can be represented as the collision of a wave with the observed object. The limit to seeing is determined by the length of the wave. This must be smaller than the object viewed, otherwise its presence, like the post in the sea, produces a negligible effect. (Photo Keystone Press Agency)

To release and manipulate the molecules of which the nut is composed, more subtle means are needed, where the energy can be put directly into the molecules so as to break the molecular bonds, rather than be dissipated over a whole volume. At the same time, our looking has to become more indirect. The microscope allows us to observe visually down to a certain limit, but if the object of interest is of the same dimensions as the wavelength of the light we are using to look, the information becomes ambiguous. A breakwater creates a dramatic change of pattern in the advancing waves, whilst a solitary post is left unheeded. We have learnt that light comes in packets, and that light of a certain wavelength is associated with packets of a certain energy; the shorter the wavelength, the higher the energy. We return again to the principle that the smaller the object we wish to study, the higher must be the concentration of energy we put in.

1/The Long Search

Research into the fundamental structure of matter has led us to understanding that crystals are composed of compounds, which themselves are composed of atoms arranged in very precise orders. At the end of the last century, the table of different atoms that had been identified looked so elegant and self-consistent, it seemed that physics was nearly at an end. But there were already anomalies appearing; the phenomenon of radioactivity, for example, did not fit into the established order. Research continued, and once more by attacking matter with projectiles of higher energy a new level was uncovered, and the essential structure of the atom was revealed. It was recognized that the atom consisted of a massive centre of diameter only about a hundred thousandth that of the atom, surrounded by protective layers, the whole system resembling in certain respects the Sun and the planets. The elements which were the basis of chemistry were characterized by the number of electric charges carried by the central nuclei, which in normal atoms were exactly counterbalanced by a similar number of oppositely charged electrons outside. Inside the nucleus, there seemed to be two sorts of particle, one carrying the electric charge, which was termed the proton, the other of similar mass, but with no charge, called the neutron. At one stroke, 92 elements had been explained in terms of just three particles, and when the characteristics of the atom had been properly quantified, the whole of chemistry and the physics of solids and liquids became comprehensible—enormously complicated, but comprehensible.

But still all was not quite well. A new particle had to be invented to allow us to keep one of our central beliefs; that is, that in any complete system, the total quantity of energy remains the same. Energy cannot be created or destroyed. (It is not so many years since this concept replaced the one that mass was constant). It was proposed that in addition to the massless particles of light which had been called photons, there was yet another massless particle which was given the name neutrino. Once again a deeper layer was showing.

Radiation emitted by naturally occurring radioactive materials had given us projectiles some 10 million times more powerful than had been available previously, and had revealed to us the structure of the atomic nucleus. We were going to need projectiles much more powerful yet, if we were going to reveal the next layer in the unknown. One natural source of projectiles is the steady shower of cosmic rays entering our atmosphere from outer space. But work with cosmic rays is slow and tedious. The atmosphere itself distorts considerably the incoming radiation, intensities are low, and the degree of planning possible in an experiment is extremely limited. These cosmic rays gave us glimpses of a new world, but this could be penetrated only by building machines able to generate particles with huge energies. Only by using the elementary particles themselves as both projectile and

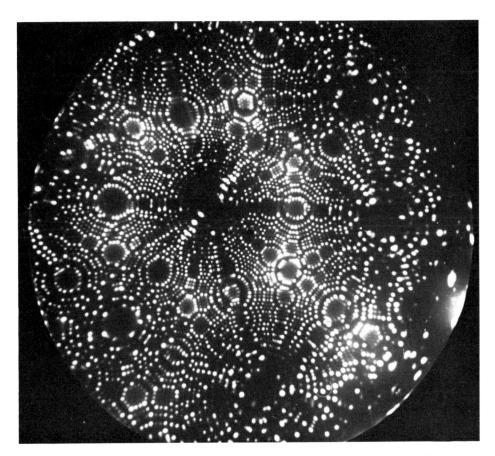

Atoms or electrons accelerated through an electric field of thousands of volts are associated with a wave of dimensions close to those of molecules and atoms. The electron microscope and field ion microscope extend the range of 'seeing' down to the level of the big atoms, but already the image needs interpretation; it is not a picture in the normal sense. (Photo T. J. Godfrey and G. D. W. Smith, Oxford)

target could the energy available in an interaction, or collision, be concentrated into a sufficiently small volume for new detail to become observable.

Two types of projectile are readily available; the proton and the electron. Protons are most easily obtained from hydrogen as the nucleus of hydrogen consists of one proton only. Electrons can be released from any element, as they are relatively loosely bound, and are given off readily by many materials when these are very hot.

1/The Long Search

An elementary accelerator is a television tube in which electrons given off from a hot source are accelerated between plates across which a high voltage is maintained. The beam of electrons is manipulated by electric and magnetic fields. (Photo Camera Press)

Either of these particles may be used. As they are electrically charged, they can be both accelerated and manipulated. If they are released between two plates connected to an electric battery, so are subject to an electric field, they will be accelerated to one or other of the plates, and the energy they acquire will be directly proportional to the voltage imposed; if the voltage is 12, then each will acquire an energy of 12 electron volts, usually written as 12 eV.

A familiar accelerator is the television tube, where the voltage imposed is a few kilovolts, and electrons released from a hot electrode are accelerated to an energy of a few thousand electron volts (keV), following which they impinge upon the screen where they produce scintillations. Pictures can be drawn because, in addition to being attracted or repelled by an electrically charged plate, i.e. accelerated in an electric field, a beam of charged particles acts like a conductor carrying an electric current. When placed in a magnetic field the conductor is subjected to a force. In the case of an electric motor, the armature turns; in the case of a beam of charged particles, the path of the particles is bent into an arc of a circle. Thus, electric fields can be used to accelerate the particles, and magnetic fields to guide them. The great disadvantage of the electron, as compared with the proton, is that because of its very small mass, when it moves along a curved path at very high energies, it loses energy continually by radiation, and there are practical limits to the energies which can be attained. Consequently, protons have been the projectiles of choice for the joint European projects, and the biggest machines elsewhere in the world. The energies these machines are designed to impart to the protons are measured in thousands of million electron volts, GeV for short.

It is impossible for so many million volts to be generated between two plates. The technique adopted is to make the particles travel in a circular orbit, and to give them modest increases in energy over and over again until the desired energy has been reached. The beam of particles is then diverted out of the accelerator and led away to a convenient area where it collides with a target. The products of interactions of these protons with the atoms of the target, in their turn, can be led away to experimental zones, sorted, and selected en route by subtle combinations of electric and magnetic fields, finally to collide with yet another target, following which the tertiary products are analysed.

All the detectors used to study what happens depend upon the action of a moving electric charge upon the environment. There is no known way of observing those particles which carry no electric charge (apart from electromagnetic radiation). Their presence can only be inferred from a careful analysis of the energy balance.

Three principal electronic detectors are in use. The scintillation counter exploits the well known phenomenon used in luminous watches and television screens, where the impact of a high-energy particle on a fluorescent material produces a flash of light. The spark chamber, which exists in a variety of forms, resembles a lightning discharge. The passage of a charged particle through a gas, across which an electric field is established, causes an electrical breakdown and a spark to jump across. The Čerenkov counter exploits the electrical analogue of the sonic boom. In the same way that Concorde, travelling at a speed higher than the speed of sound in the atmosphere, produces a shock wave of sound in the form of a cone spreading out from the nose, so, a high energy particle travelling through a

1/*The Long Search*

 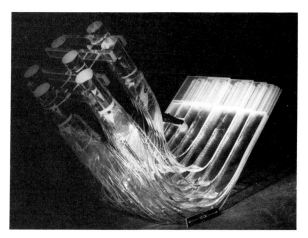

The scintillation counter, like the luminous paint on a watch dial, exploits the capacity of some materials to emit flashes of light when bombarded with energetic particles. In the watch, only the overall glow is important, but in the scintillation counter, the individual flashes are piped down light guides to sensitive electronic detectors which measure the size of each flash.

transparent medium at a speed faster than the velocity of light in that medium, produces a shock wave of light spreading out along the trajectory of the particle. The angle of the cone of light is a measure of the speed of the particle.

A quite different detector is the bubble chamber, which is a container in which a volume of liquid (for example, liquid hydrogen) is held under pressure just below its boiling temperature. When the pressure is suddenly released, the liquid wants to boil, beginning along local irregularities, such as the ionized track of a charged particle passing through. The lines of bubbles which form can then be photographed, and the liquid re-compressed before bulk boiling takes place. The physicist obtains a permanent photograph of the tracks the particles have followed.

The amount of information obtainable from any one set of collisions is small. The electronic equipment can detect when particles pass through, what their tracks are, and perhaps their speed. If a magnetic field is interposed, their probable masses can be deduced from the curvature of the tracks. From the bubble chamber pictures, the physicist has a detailed photograph of the line the particles travel, but very little else. It must be remembered that only the charged particles leave tracks, and every event is unique and unrepeatable. Only by the statistical analysis of thousands, even hundreds of thousands, of similar events, can the characteristics of the participants be deduced.

1/The Long Search

In the spark chamber, conditions are created similar to those naturally occurring in thundery weather. A lightning flash produced between two planes of stretched wires, follows the passage of a high energy particle whose position is thus recorded. (Photos Camera Press and CERN)

As the various layers of information on the structure of matter were uncovered, so it was possible to say that molecules were composed of atoms, and atoms were made of nuclei surrounded by electrons. When, however, the research was pushed one stage further, some of the simple expressions such as 'made of' and 'composed of' began to lose their meaning. It was discovered that when two protons collide, far from the result being pieces of proton, new particles are formed, and the two protons are often left intact. The new particles have masses, for example, of constant and definite value, but they have proved to be more numerous than could ever have been anticipated. The total score at present is of the order of two

The Čerenkov counter is the electrical analogue of a supersonic plane detector. Concorde travelling at a speed faster than sound produces a sonic boom, while a particle travelling through a medium at a speed greater than the speed of light in that medium produces a cone of light. Complex optical systems, sensitive to a particular cone, detect the passage of particles moving at a particular speed. (Photos Camera Press and CERN)

1/The Long Search

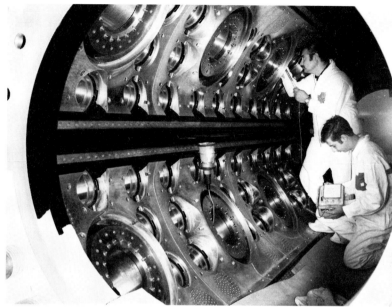

The bubble chamber was inspired by the observation of the way bubbles formed at particular places when a beer bottle was opened. Inside a bubble chamber, a liquid is maintained under pressure just below its boiling point. When the pressure is suddenly reduced, bubbles form along the tracks of particles passing through. These tracks are then photographed and the liquid re-compressed. (Photos Camera Press and CERN)

hundred. Their life-times vary from the permanence of the proton to appearances of as short as a millionth of a millionth of a millionth of a millionth of a second. So the work has gone on to try to classify them, to put them into families, to determine the genus, the species, and the varieties, and to pick out the salient characteristics which are their intrinsic qualities.

There seem to be two main families: that of the proton, the hadrons, is very large, and that of the electron, the leptons, rather restricted. Attempts to tidy up the electron family have met with little success. One of the great mysteries of nature is, for example, why there should be a particle, called a muon, two hundred times as heavy as the electron and yet from all other points of view, apart from its life-time, identical to it. On the other hand, the proton family of particles has been found to form a very neatly symmetrical pattern, well represented by groups of 8, 9 and 10 particles. This tidy arrangement does not in itself explain how the pattern arises. But in a rather similar way to the earlier explanation of the neat pattern of properties of the chemical elements by the discovery of the proton, neutron and

Europe's Giant Accelerator

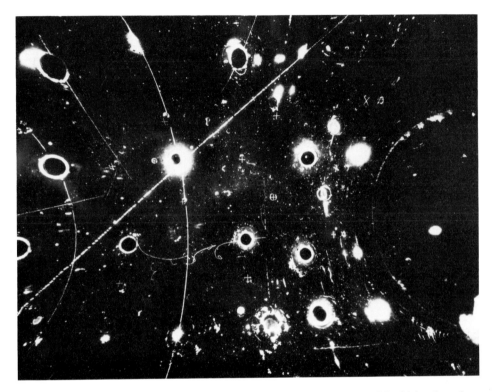

A photograph of particle tracks taken in the CERN heavy liquid bubble chamber, Gargamelle. The fine traces a little below and to the left of centre show one of the first events ever discovered in which a neutrino collided with an electron without other particles being produced. A few pictures like this, out of hundreds of thousands taken, suggest that two of the four forces known in Nature may be different aspects of a single force.

electron, it has been proposed that the hadrons can all be derived by combinations of hypothetical component particles which have been given the name quarks. With new concepts, there is always a problem of language, and while quarks were whimsically named after James Joyce's "Three quarks for Musther Mark", the fourth member of the group which has had to be introduced is designated by its special property of 'charm', a word that has no connection with its normal usage, but reflects perhaps the physicist's search for beauty and elegance in his description of Nature.

1/The Long Search

Only by the careful measurement and analysis of the tracks produced can new discoveries be made. In a single run of one experiment, hundreds of thousands of pictures are taken. The process of measurement is becoming progressively automated as new machines, controlled by computers, are developed.

In the early days of the big modern accelerators, the discovery of a new particle, or the identification of the characteristics of another, had seemed almost a daily occurrence, but the 60s saw a pause in the torrent of information, accompanied by increasing perplexity as to how to interpret it. The late 60s and the early 70s were lean years in high energy physics, until suddenly a new spate of results brought in by bigger machines and bigger detectors began to transform the scene once more. Size, once again, was the key to probing deeper inside the tiniest particles of Nature, confirming the view than an accelerator of yet higher energy was needed to further our understanding of the fundamental properties of the sub-nuclear world.

It is now definitely established that the proton has a structure: that it contains three point-like objects which seem loosely bound together by some sort of glue, rather like a grape with three pips. But what the skin of the grape is remains a mystery. Studies at very high energies should reveal the nature of the objects and why they hang together as they do.

For 40 years, four fundamental forces have been recognized in nature. These are the familiar ones of gravity (controlling the movement of the heavenly bodies and our movements here on Earth), electromagnetism (controlling chemistry and the physics of solids), and two very short-range forces, the weak and the strong, which are in evidence only over distances the size of the nucleus. One of the deepest ambitions of physicists is to reduce this number. In the same way that electricity and magnetism were, at the end of the last century, shown to be two facets of the same force, so it is hoped that the four forces currently recognized can be shown to be manifestations of one central force. In the early 70s, the first indications came that this may not be an idle dream, as a relationship between the electromagnetic force and the weak nuclear force was uncovered.

The hadrons are affected by the strong force whereas the leptons are not. So when leptons interact with each other, it is principally the weak force that we see at work, although if a particle carries an electric charge, it will also be affected by the electromagnetic force, which is the one responsible for most everyday physical and chemical phenomena. Similarly, only particles with mass are influenced by the gravitational force. The neutrino, having neither charge nor mass, hardly ever interacts with anything, and can pass straight through the Earth as if it were not there. As it is affected only by the weak force, it is ideal for studying this force, but the rarity of events in which it is involved means that a huge number of neutrinos must be generated to get enough of these events. One of the uses of Europe's new accelerator is to provide neutrinos in quantities never previously available, either naturally or from lower energy accelerators. At the same time, large numbers of muons (the heavy electrons) will be formed, giving opportunities further to unravel the mysteries presented by these particles.

After many years of painstaking cataloguing of new particle data, and precise and accurate measurement of phenomena which may or may not be significant, it is possible that we are on the threshold of a new great step forward in knowledge. It may require conceptual leaps as important as those of the early part of this century. We may find that the very idea of building things out of components at this level has to be rejected, that the real stuff of matter is not to be found in particles themselves, so much as in qualities which are revealed only in the context of a system where the individual owes its identity to the society in which it finds itself, as much as the identity of the society is determined by its individual components. High energy physics, intellectually, may not be so far removed from the social problems of our time as might at first sight seem probable.

2/The Practical Dreamers

In the late 1940s war-emergent Europe was riven by the animosities between former allies (as expressed in the 'cold war'). We were seeking to adapt to living in the threat of the atomic bomb, and the restricting hand of military secrecy lay over much of the advanced work then being done in science. Whilst a group of inspired politicians was searching for the means to re-establish European culture, some leading physicists had a vision of a new European co-operation in physics, of a research organization freed from the restrictions of oppressive security barriers, able once more to do research in the traditional open spirit. Their ideas married well with the aspirations of the politicians; a partnership was forged between scientists and government representatives, which was to become the solid base upon which a completely new institution could be created, one which has come to be regarded as a model of European collaboration.

In the early discussions, first given voice in a message from the distinguished French Nobel laureate, Louis de Broglie, to the European Cultural Conference in Lausanne in December, 1949, nuclear physics was not even mentioned, but it quickly became identified with the new initiative. Pierre Auger as UNESCO's Director of the Department of Natural Sciences, organized the movement, and through funds voted first by the Italian Government and soon afterwards by the French and Belgian Governments, it was possible for UNESCO in 1951 to call together an inter-governmental conference of European states to consider the creation of a new international research institution. This institution, which came to be known as CERN, would be equipped with research facilities beyond the scope of the individual partners, and able to compete with the best that might be built anywhere else in the world.

This was a period when men of science had become accustomed to walking in the corridors of power, to having the ear of governments and to bearing the responsibility that matched the influence they wielded. The people who were associated with CERN in its beginnings were the great names of the day: scientists such as P. M. S. Blackett and J. D. Cockcroft from the UK; P. Auger, L. Kowarski, and F. Perrin from France; E. Amaldi and G. Bernardini from Italy; N. Bohr from

Denmark; W. Heisenberg from Germany; and P. Scherrer from Switzerland. Their influence, however, would not have been sufficient without the drive and inspiration of the men of government committed to the European image. Men such as R. Dautry and F. de Rose, in France, J. Willems in Belgium, J. H. Bannier in Holland, Sir Ben Lockspeiser in the UK, and A. Picot in Switzerland. Picot was the person who swung the Canton and City of Geneva to agree to the installation of this new institution on its territory, at a time when the word 'nuclear' spelt war, destruction, and the dangers of radiation.

The movement was not lacking in friends abroad, and the benevolent help of the United States was strongly in evidence, in the persons particularly of I. Rabi and J. R. Oppenheimer. From the earliest beginnings, a special brand of co-operation and rivalry was developed between Europe and the United States, which has become stronger with the years whilst remaining always on an informal basis. There is a tendency to belittle the seriousness of American research and innovation, whilst at the same time revering it to the point where only American confirmation of a new result can put the seal on its authenticity. This is a love–hate relationship in which the qualification of BTA (Been to America) was at times more powerful than the more academic qualifications of European institutions. But the partnership has developed to the point where in the United States the qualification BTC (Been to CERN) is becoming almost as important, and for many years now the rivalry across the Atlantic has been between equals, and the exchange of confidence great.

Co-operation with the Soviet Union was to come much later. When it began it was one of the first breaks in the 'cold war', and high energy physics has continued to be an area in which we can talk of a truly world scientific community.

In the early days, the position of the UK was ambiguous. When, at the second meeting called by UNESCO and held in February 1952 in Geneva, a convention was signed setting up a provisional organization called the Conseil Européen pour la Recherche Nucléaire (CERN), eleven Governments signed the convention, but the UK remained as observer. Characteristic of the period, UK participation was marked by personal drive and initiative, coupled with political timidity. Although unwilling to commit the country to a straightforward membership, the UK made an *ex gratia* payment of an amount equivalent to what its formal contribution

Opposite
P. Auger (**left**), *of France, and Nobel Prize winner Niels Bohr, of Denmark, discuss the plan to set up a new European laboratory for the study of the fundamental structure of matter.*

In February 1952, at a meeting organized by UNESCO, the convention establishing the provisional organization, CERN, is signed. With Bohr are (**right**) *P. Scherrer, of Switzerland and A. Picot who was the chief instrument in obtaining the agreement of the Geneva authorities to accept the laboratory on its territory.*

would have been, and in the work of the Conseil under its Secretary-General, Professor Edoardo Amaldi, British delegates played an active role.

By July of the following year, the Conseil had been able to elaborate the convention which would establish the European Organization for Nuclear Research, and by September 1954, a sufficient number of states had ratified this convention, to guarantee at least 75 per cent of the financial commitment. The UK had made amends by being almost the first to ratify.

So the provisional Conseil was dissolved, and the European Organization for Nuclear Research formally came into being. Already, however, the acronym, CERN had become established, and ever since has remained the title by which the Organization is known all over the world. The founding member states were:

2/The Practical Dreamers

Above left
When the Convention setting up the Conseil Européen pour la Recherche Nucléaire is signed in February 1952, E. Amaldi (left) of Italy is appointed Secretary-General. With him is L. Kowarski, of France, who became head of the scientific and technical services division in the new laboratory and stayed at CERN until his retirement.

Above right
G. Bernardini, another Italian with a mission to promote European cooperation. Later he was to become the first president of the European Physical Society which held its inaugural meeting at CERN in 1968.

Belgium, Denmark, Federal Republic of Germany, France, Greece, Italy, the Netherlands, Norway, Sweden, Switzerland, the UK and Yugoslavia. Later, Austria and Spain were to join the Organization, but subsequently Yugoslavia and then Spain withdrew because of financial difficulties.

The convention to which the States had agreed has been a determining factor in the dynamic development of the Organization. It has never been used by its members to try to manipulate the Organization to national ends, and at CERN

Europe's Giant Accelerator

Opposite, top
*First president of the CERN Council was Sir Ben Lockspeiser of the UK who is here listening to a report from the first Director-General, Nobel Prizewinner Felix Bloch, of Switzerland. In the background (**right**) is a young man who was to make an immense contribution to the success of the Laboratory, as Director of the Proton Synchroton Division, Director-General of CERN (1960–1961), Director-General of the SPS and then Executive Director-General of CERN since January 1976—John Bertram Adams.*

Above left
*At the sixth meeting of the CERN Council held in July 1953, a convention was signed which set down the terms of reference of the permanent organization—the European Organization for Nuclear Research. Prominent among the statesmen who had guided the project through was J. Willems (**foreground**) of Belgium.*

Above right
Another prominent statesman with a continuous association with CERN to the present day is J. H. Bannier, of the Netherlands. In the background can be seen C. J. Bakker, also of the Netherlands, who was Director-General of CERN from 1955 to 1960.

*Scientific representative of the UK to the Council in the early days was Sir John Cockcroft (**second right**), whilst on the far left can be seen Gösta Funke, the delegate from Sweden, whose tact and diplomatic skill were to be of crucial importance when, as president of Council in the difficult years of 1967–1969, he safely steered the Council through many potentially inflammatory sessions.*

Europe's Giant Accelerator

itself although copies are freely available, they are rarely to be found in evidence. The convention has the huge merit of being an open-ended document, designed to facilitate the objectives of the Organization rather than limit them. The objectives are set out in wide terms and widely interpreted; the means are clearly and explicitly defined.

The aims are defined as follows:

The Organization "shall provide for collaboration among European states in nuclear research of a pure scientific and fundamental character, and in research essentially related thereto." In the same clause, the convention speaks of the organization and sponsoring of international co-operation in nuclear research and the widest interpretation has always been placed on this phrase with the full agreement of the member states. To begin with, the means were defined as the construction and operation of two particle accelerators a synchro-cyclotron with an energy of 600 million electron volts (MeV), and a proton synchrotron of energy above 10 thousand million electron volts (GeV), this being the size of the largest machine then under construction (in the USSR).

The convention continues with other provisions exceptional in their openness, but which have stood the test of time. Regardless of the size of their contributions, all member states are equal in their voting rights, and the only questions which require unanimity are those which affect the fundamental structure of the organization, or change commitments previously entered into. Many of the decisions are taken by simple majority, some by two-thirds majority, with the result that delegates are rarely to be found taking decisions on narrow majorities, or seeking to oppose the obviously general will. Virtually all decisions of importance are taken by consensus, working out a formula to which eventually all member states can subscribe.

There are no quotas concerning either staff or contracts which depend on the members' contributions, and, indeed, governments have virtually no rights in the exploitation of the facilities of the Organization. The Directors-General have total responsibility for everything that goes on within the Organization, and participation by member states is through their institutions and individuals. Legalistic obligation has given way to intelligent implementation, and friction avoided by seeking a reasonably fair distribution between the states over the years, where this has been possible, even if there may be imbalances at any particular moment. Nonetheless, countries who are not interested have had no obligation to use the facilities, whereas those with a will to make the most of what the Organization offers have been able to take advantage. On the financial side, payment is virtually automatic being based upon a scale related to the average net national revenue of

2/The Practical Dreamers

each member state, a scale which is updated every three years. For many years, also, a four-year rolling budget system has been adopted with largely automatic indexing of these budgets according to a complex calculation. The method of calculation has involved long debate and the indexing has not always been fully applied, but the arguments in the end have been about the odd 1 per cent, and CERN has been able to plan its forward programmes in the confidence that budgets finally allocated would be close in spending power to those that had been anticipated some years previously.

The convention states that not only shall the Organization have no concern with work for military requirements, but also that the results of its experimental and theoretical work shall be published, or otherwise made generally available. In consequence, the whole nonsense of commercial secrecy which has bdeevilled the operations of so many atomic energy commissions has been avoided. Not only has this simplified procedures, but also it has resulted in commercial benefits to contracting companies far in excess of that produced by the operation of patents or special access agreements.

The cynic would suggest that such an intelligent convention could be drawn up and signed only because CERN started off as a small and insignificant organization, whose only usefulness was to act as an insurance policy against the discovery of some physical phenomenon, which in the long term might lead to a new source of power. For a few years, it may be that some politicians confused the basic and fundamental nature of the research at CERN with the more academic aspects of the development of nuclear power, but this did not last long, and by the time CERN was seen to be an organization of growing significance, operating on a scale which was no longer negligible, it had already established a reputation for honesty, efficiency, and good book-keeping. The honesty is evident in two respects: when CERN presents a proposal to its member states it is not a proposal for half a project, assuming that once half-way committed the member states could hardly refuse the other half. It has always recognized that a machine on its own is of no value; it is there to be used, and its exploitation needs equipment of comparable complexity to the machine itself. In addition, it has known how to do its costing, and it has not been prepared to bargain away a position of realism by accepting cuts essentially imposed by the horse-dealing tendencies of politicians. As such it has been respected by the member governments, its costings are expected to be realistic, and adhering to programmes and budgets no longer occasions surprise.

The staff of CERN who began to assemble in Copenhagen, and then in Geneva, had no promise for the future, no certainty that they would have continuing employment. But their willingness to stake their careers in the promise of the idea meant that the project to build CERN was already off the ground before the formal agreement had been given by the member states. It meant, also, that those who had been attracted to the project were not the faint-hearted, but the adventurous physicists and engineers driven often by a conviction of the need for European co-operation, and ready to take the risk of putting it to the test. As with so many new organizations, it was a cheerful informal company that got down to the difficult grind of designing and constructing the machines. Since that time, growing size has brought with it the need for an administration that is more formal, rules and regulations which are set down carefully on paper, contract terms which are immutable and sometimes inhuman. Nevertheless, when the administration of CERN is compared with national establishments, or with similar centres in the United States, in spite of the heavier load required to cope with the multilingual problem, CERN's cost effectiveness record bears comparison with the best. The general lines of policy are established by the CERN Council which consists of two delegates per member state, and tradition requires that one of these delegates is a physicist and the other an administrator, so preserving the dual partnership which has proved to be of such value. Much of the hard work and supervision of day-to-day business is done by a Finance committee, composed of one delegate per member state. Both Council and Finance committees have benefited from the continuity of membership of many of the national delegates, whose identification with the aims of the Organization has resulted in their being not simply a representative of their member state at CERN, but also a proponent of CERN in the member state.

Council is also advised by a Scientific Policy Committee made up of eminent scientists. This is the most difficult of the committees to keep young and vital, and with the passing years it found itself not fully equipped to cope with the complexities of the big new project, nor the harsh decisions needed when budgets levelled off and priorities were savagely contested.

International organizations which require for their main decisions the agreement of a dozen countries have an in-built inertia which makes decision-making slow. Nevertheless, this inertia is compensated by stability, which may over the long run be of more importance than a capacity for making rapid changes in policy.

Opposite
Even before the Convention setting up the European Organization for Nuclear Research had been ratified by a sufficient number of states for it to have legal status, work on the new Laboratory started. Within a year, buildings were already taking shape and on 10 June 1955, Felix Bloch laid the foundation stone.

2/The Practical Dreamers

Europe's Giant Accelerator

*By the beginning of 1956 the form of the Laboratory was taking shape on the Franco-Swiss frontier—an irregular line to the left of the civil engineering works. The building for the 600 MeV synchro-cyclotron (**upper right**) is largely complete as work goes on preparing the 200 m diameter structure (**lower centre**) which will house the PS.*

This very inertia spared Europe the savage cuts to the high energy physics programme that were made in the United States in the early 70s and which, for lack of confidence, would surely have been mirrored in Europe if there had been no mandatory pause for reflection.

The real danger though to an ageing scientific organization lies not in the growth of the number of permanent civil servants, in the complexity of the administration, and in the accumulation of out-of-date half-forgotten projects, but in the increasing weight of the scientific establishment and in too strong a compulsion to exploit to the full the availability of the existing machines. Many a factory finds its work programme being defined by the machine tools it has available rather than by a rational evaluation of what it can do most profitably to meet the existing market.

2/*The Practical Dreamers*

International scientific recognition of the importance of CERN came in 1958 when CERN was host to the International Conference on High Energy Physics attended by ten Nobel Prize winners. In this picture are, from left to right: C. Powell (UK); I. Rabi (USA) who had personally made an immense contribution to the negotiations leading to the establishment of the Organization; W. Heisenberg (Germany); E. M. McMillan, E. Segré, T. D. Lee, C. N. Yang and M. Goldhaber (all of the USA).

Similarly, with a scientific establishment, there is a tendency to settle for experiments which suit the machines, rather than those which will possibly further the subject the most. Europe has a tendency to establish hierarchies, and an ageing scientific establishment will prefer the conventional proposal of a reliable man to the outlandish suggestions of a young unknown. The scientific establishment has a strong tendency to prejudge not only what should be discovered, but also who should discover it.

This tendency has been countered at CERN by a system of open committees in which scientists proposing an experiment must fight for its acceptance before their

The world's first strong focusing proton synchrotron achieved its design energy of 24 GeV on 24 November 1959. During the commissioning, John Adams speaks to the central control room, while a group of the builders watch anxiously what is happening on a cathode ray tube. Directly in front of the screen is Wolfgang Schnell, and in the foreground centre, Ch. Schmeltzer.

peers and competitors, before it then can be considered as a real contender for time on one of the machine programmes. This, of course, militates to a certain extent against the unconventional proposal and favours the experienced group leader with an established position and a talent for committee operation. Still, the openness of the discussions prevents serious abuse, and there is no harm in developing a core of bright scientists with tactical experience.

CERN was set up to do research at the frontiers of what was physically possible. It was an institution at the limits of what was politically conceivable, and started off by matching its technology to these high ambitions. Accelerators up to the early 50s had been built on the philosophy that the particles to be accelerated had

2/The Practical Dreamers

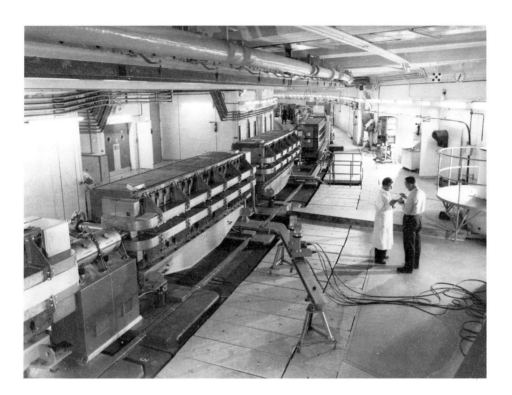

The first experiment is mounted at the CERN PS, a modest group of two detectors inside the ring of the accelerator. As the years went by, the size and complexity of the experiments were to increase steadily; the first experiments on the SPS are bigger and more complicated still.

to be handled gently and coaxed in a big vacuum chamber to make their rounds of the machine, oscillating gently up-and-down and in-and-out, under the influence of a gently varying magnetic field. When the design of the big CERN machine was about to be frozen, news came from the United States of a theoretical study that suggested that if they were treated roughly and forced—first this way and then that—the results could be a fine beam, much as in creating a thin pencil of plasticine by squeezing and rolling it. Forthwith, the decision was taken to scrap the existing design, and CERN embarked on a huge experiment of building the world's first strong-focusing proton synchrotron with a design energy of 24 GeV, an approach that came to be adopted by all accelerator builders from then on. A new generation of machines was born

At a press conference held at the inauguration of the PS, the platform consisted of, from right to left; J. R. Oppenheimer, E. M. McMillan (both of the USA), N. Bohr (Denmark), J. B. Adams, F. de Rose (France), C. J. Bakker, J. D. Cockcroft (UK), S. M. S. Balke (Germany) and, at the extreme left, F. Perrin (France).

In essence the modern synchrotron consists of a nearly circular ring of magnets, through which passes an evacuated tube of small cross-section. In the tube circulate protons (produced by stripping the outer electrons from hydrogen atoms) which have been given an initial acceleration in a series of accelerators of different characteristics before being injected as a pulse into the synchrotron ring. There, the magnets guide and focus them into a thin beam while r.f. fields split up the ribbon into bunches that are then accelerated up to the desired energy. As the energy increases, the field in the magnets is increased in synchronism to keep the orbit constant, and the frequency of the accelerating field is synchronized to the orbiting period. When the top energy is reached, the protons are extracted from the machine and the magnets are de-energized, to make them ready to receive the next injection pulse.

2/*The Practical Dreamers*

From the beginning, the relations between CERN and the USA laboratories had been very close, while contacts with the Joint Institute for Nuclear Research in the USSR developed during the 1960s. Following an agreement with the relevant State Committee of the USSR in 1967, equipment was shipped from CERN to the 76 GeV synchrotron at Serpukhov for joint experiments to be carried out there.

Great accuracy and stability are needed if the protons are not to be lost during their many turns of the machine, and at CERN the pattern was set of requiring the highest possible standards in engineering, with the determination not to accept second quality, preferring to abandon a project rather than do it other than excellently.

It was in the early days of the building of the CERN proton synchrotron (PS) that a new name came into prominence; John Adams, a young protégé of Sir John Cockcroft, whose position, as Member for Research of the UK Atomic Energy Authority, he was subsequently to take after being Director of the Culham Fusion Laboratory. Adams, whose brilliance as an engineer was matched by a natural capacity for leadership, became leader of the design team of the CERN PS

J. B. Adams was succeeded as Director-General of CERN in 1961 by V. K. Weisskopf (USA), (left), and he in his turn in 1966 by B. P. Gregory (France), (right). Between them stands L. Van Hove (Belgium), who became Research Director-General of CERN in 1976.

and responsible for its construction. He was made Director-General of CERN soon after the PS came into operation, in place of Professor C. J. Bakker who was killed in an aeroplane crash. Having been largely responsible for the institution of the committees which controlled the operation of the CERN PS, Adams then returned to England in 1961 while his old associates got down to thinking of the next stage.

CERN had by this time well learnt the value of consultation with the scientific community in its member states, and before going too far, a new committee—the European Committee for Future Accelerators (ECFA)—was set up under Amaldi, with the job of assembling physicists from the European countries with those of CERN to consider what the next stage should be. This Committee submitted a report to the CERN Council in 1963, which recommended a pyramid structure of machines. At the top would be the international facilities consisting of a CERN

2/The Practical Dreamers

With the agreement in December 1965 to build the Intersecting Storage Rings (ISR) on a site in France adjacent to the existing laboratory, CERN became the first international centre to straddle a frontier. When the administrative problems were resolved, G. H. Hampton, director of administration at CERN, in the presence of representatives from the local authorities, cuts the ribbon that symbolically divided the two halves of the Laboratory.

enlarged by adding storage rings to the existing PS, and a new laboratory based on a proton synchrotron ten times the size of the existing machine. At the base of the pyramid should be the national centres and university facilities and in between the extremes, a series of regional machines. CERN was anxious to reassure, in particular, the national laboratories, that the international centre was not competitive with their activities, but a logical projection of their work which would maintain the flow of students and new physicists, and incidentally funds. The logic was inescapable: the more put into the international centre, the more correspondingly must go into the national activities. This mood which seems almost unbelievably lavish now, lasted through into the late 1960s.

The big new machine which was proposed would be the principal instrument of a completely new high energy physics laboratory to be built somewhere in Europe, not in Geneva. The accelerator would have a diameter of 2·4 km and would be equipped with a number of tangential beam lines stretching away like the rays of a catherine wheel. The estimated cost at 1966 prices was 1700 million Swiss francs (Sw. Fr.), and it was proposed that site work should start in 1969 for the machine to be operational in 1976, a date that seemed more and more unattainable as the years passed.

The proposal to build intersecting storage rings (ISR) was in many ways more controversial. In traditional physics, the accelerated beams are directed to solid targets, where beams of secondary particles are formed which are led away to the experimental areas. What was proposed with the ISR was that beams of particles already accelerated in the PS should be stored in two intersecting rings where they would travel in opposite directions. At the intersections they would meet in head-on collision. It was not sufficient to use the output from one acceleration cycle of the PS to fill each ring as the number of particles was such that few collisions would be produced. To have any reasonable rate of interaction, it was going to be necessary to store hundreds of pulses from the PS in each of the rings, stack them side by side, and then maintain them in their orbit for periods as long as a day. British physicists were openly sceptical that the ferocious technological problems could be mastered; they were more anxious that the big new synchrotron should be built. If everyone had appreciated fully the real magnitude of the technological problems, it is doubtful whether such a hazardous project would have been embarked upon. As it was, the specification looked to be at the limit of what was feasible. It was certainly not appreciated that only by going an order of magnitude in pressure beyond what was contemplated, in a vacuum system 2 km round, would it be possible to store currents of the desired intensity.

The machine builders, though, had their priorities clear. This was the machine they wanted to build at that time; it was challenging; it could be built alongside the existing machines so, not only were the political problems easier, but it meant that families which had become established in Geneva would not have to move. During 1964, the two projects—one for the ISR, and the other for a machine, now being called the 300 GeV, were pursued in parallel. An international discussion in Vienna raised the question of whether the time had not come for a world machine to be built, but the USSR was building its 70 GeV machine at Serpukov, and the USA was already actively planning a proton synchrotron of 200 GeV. The conclusion was that below 1000 GeV, the separate continental blocks should go it alone.

2/The Practical Dreamers

In March 1971, the ISR construction project was deemed complete and experiments were installed at the intersecting points around the ring to study collisions between protons at the highest energy ever available. This project was completed four months ahead of schedule and within the budget estimate prepared six years earlier.

It was then that the first serious site proposals for the new accelerator were sought, and the member states were given as requirement a minimum area of 20 km², with the minimum width of 3·5 km. The site had to be geologically stable, able to accommodate a tunnel in which the accelerator would be constructed and with access to large quantities of cooling water and electric power. It had also to be situated in a region readily accessible from all over Europe and capable of attracting the very best people. There was still a great deal to do on the project, and Council in December 1965 gave the go-ahead to build the ISR on a site adjacent to CERN offered by the French whilst instructing the Organization to pursue its study of the 300 GeV machine. The result of this decision was that CERN doubled its area from about 40 to 80 hectares, and became the first international organization to straddle a frontier. The fence around the site was changed so that access for personnel was only possible through Switzerland, but provision was made for goods arriving from France to by-pass the multiple Customs operation that would otherwise have been necessary. Minor problems arose in operating this novel situation; for example, the problem of the coffee machines on the new site which could not benefit from Customs franchise as they were not connected with the work of CERN. They were consequently subject to French law, but the coffee could arrive only from Switzerland. From the technical side though the arrangement worked well, and in a very short time people were forgetting they were crossing a frontier as they moved from the PS to the ISR building site.

The ISR project budget had been agreed at 332 million Sw. Fr. and the date for operation was set as July 1971. It was a source of satisfaction to scientists and administrators alike that the machine became operational in March 1971, and the gross budget, corrected to the 1965 figures, added up to 326 million Sw. Fr., 2 per cent less than estimated seven years before.

The leading CERN staff who had directed both the design studies for the ISR project and the initial studies for the 300 GeV machine were Kjell Johnsen and Cornelius Zilverschoon. When the ISR project was agreed by the CERN council in 1965, Johnsen was appointed Director of the project and Zilverschoon his deputy. Despite these heavy responsibilities, they both maintained a keen interest in the 300 GeV project, which continued even after Adams took over in 1969.

3/Site for Agreement

Once the ISR project had been agreed and the construction was under way, the physics community could turn its full attention to the other project that had been urged upon Governments by ECFA in its original report: the construction of a proton synchrotron 10 times bigger than the existing CERN machine, although the number of people who would from now on be at CERN engaged actively on the project was even smaller than before. Its further development depended on the co-operation of physicists throughout the member states.

For some time, in the member states, studies had been conducted on possible sites which could house such a huge machine and its associated experimental facilities. Over a hundred sites had been considered, and these were whittled down to 22 on the grounds that some had insufficient area; others were geologically unfavourable; and some posed operational problems such as difficulty of communications. But even 22 was rather more than could be handled by the Study Group that was set up, and member states were asked to single out in their own countries what they regarded as their most advantageous proposal. This was easier said than done, and when, finally, the hard core of 12 sites was determined three were in Germany and two in Italy. Rivalry within these two countries was such that the national authorities had found it impossible to choose between them. Norway, Denmark, the Netherlands, and Switzerland were not offering sites, and they were therefore placed in the unenviable position of being arbiters in what was becoming a very keen competition.

ECFA had been re-convened at the beginning of 1966 to reconsider the validity of the original proposals, and to take further the study of what would be done with the machine once it was built. Throughout the project the physics community remembered that it was not simply a machine that was being built as an exercise in its own right, but an experimental tool which would need ancillary facilities and experimental equipment of correspondingly large dimensions. EFCA's full report was published in May 1967, and was accompanied by three huge documents which indicated that there would be no shortage of experimental proposals. There was no shortage either of ambitious ideas, such as, for example, a bubble chamber

3/Site for Agreement

which would contain 200 tons of liquid hydrogen. As liquid hydrogen has a density about 1/14th of that of water, this would mean a chamber containing three million litres!

They were heady times indeed, and few physicists thought it excessive to be demanding not only this new laboratory, which at the up-dated prices would cost 1776 million Sw. Fr., but also urging the member states to consider the construction of smaller regional accelerators, and generally to upgrade the level of expenditure in the national establishments.

The unanimity of opinion amongst the physics community, particularly as represented by ECFA, was a powerful force in its own right, and there was really only one dissenting centre of opinion. This was Nobel prize winner Heisenberg, in Germany, who was sceptical about the value of another proton synchrotron, although generally favourable to the peep at higher energies which would be afforded by the intersecting storage rings. Heisenberg, one of the great theoreticians in physics, was still hopeful of ante-dating its operation with an all-embracing theory which would synthesize the vast amount of data that had been accumulating over the previous decade. (He died in 1976 without achieving it, just a few months before the new accelerator achieved its target energy.) The position of Germany was thus ambiguous, because although the majority of high energy physicists in Germany were in favour of the project, Heisenberg's influence at government levels was considerable.

Physicists in the United Kingdom were more worried about their ability to provide enough people to do all the work that would be needed on and with the new machines, as well as keeping a full activity in the national laboratories. They, like the members of ECFA and the Directorate of CERN, were obsessed with the notion of keeping the national activity vigorous and not concentrating too much in the international centres. To ensure that CERN would not become an ivory tower, the practice had been growing of encouraging a regular turnover of

Opposite

After more than 100 sites across Europe had been studied, the number offered for the new Laboratory had been whittled down to nine by June 1967. Only four of the then thirteen member states of CERN (Spain withdrew the following year for financial reasons) were not proposing a site—Denmark, the Netherlands, Norway and Switzerland. All the other countries were actively competing to bring the Project on to their territory.

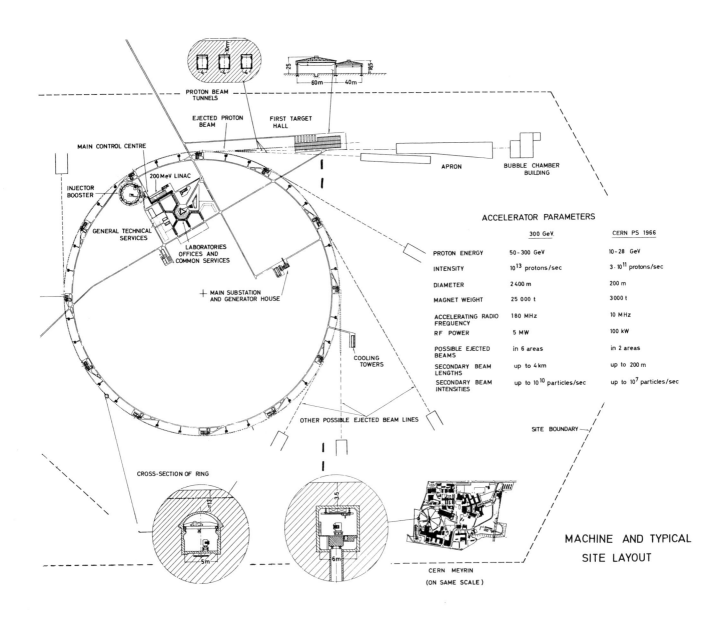

The 1964 design was of generous concept: a ring building 2·4 km diameter of domed section 5 m across and 5 m high, little different from the section of the PS. Traditional magnets were proposed which would have had a total weight of 25 000 tons, almost double the weight of the magnets eventually used. Several separate beam lines were proposed requiring a minimum site area of 20 km².

research physicists to the point where the vast majority of the research was done by teams visiting CERN for relatively short periods. The United Kingdom, loaded with its two large national laboratories—Rutherford with the 7 GeV PS and Daresbury with the 4 GeV electron synchrotron—and suffering from the traditional suspicion that anything done on the Continent could only be regarded as entertainment, made poor use of the CERN facilities. It was not until Brian Flowers was appointed head of the Science Research Council that UK physicists began to exploit in full the international opportunities for which the country was paying.

Nevertheless, it was the UK which first raised the question whether it was really necessary to have, in addition to the big international centres, national centres of comparable strength. This marked a new phase in the thinking towards CERN which found itself in the unexpected position for an international organization, of being asked to take more responsibility and a larger share of the overall funding. The UK's financial position was worsening, and it was the first country to worry about the longer-term implications of this ever increasing call on the public purse. Elsewhere, this was less in evidence, and France and Germany were still talking about a collaborative venture for a proton synchrotron of perhaps 50 GeV—although the word 'collaborative' at times seemed to have a different meaning on each side of the Rhine.

By the end of 1967, France, which had been the first to say yes to the 300 GeV proposal, had been joined by Belgium and Austria, but no-one else—and eyes were on the United Kingdom. At Westminster, the government having glumly accepted the inevitability of devaluation was looking for projects which seemed to offer a direct financial benefit, and were in consequence less and less enamoured of foreign expenditures with no obvious returns. On the other hand, they had received a report from the Council for Scientific Policy supporting the Science Research Council's view that the 300 GeV project was important not only for British but also for European basic science, even if a minority comment by two members of the Council did express concern that it would mean a very high proportion of the Council's budget for some years to come being devoted to physics.

In the neutral corner was an extraordinary report by J. N. Wolfe, Professor of Economics, and A. J. Youngson, Professor of Political Economy, in the University of Edinburgh. Starting off with some snide comments about how unlikely it was that the accelerator would be built for anything near the price quoted, on the basis of the experience of the North of Scotland Hydro-electric Board on work "not dissimilar in character", they attempted to analyse the effect on the UK's economy of siting the project in the UK or abroad. They reached the conclusion that there

Ministers of the member states were invited to CERN to see for themselves the success of the Organization. V. K. Weisskopf (left) and B. P. Gregory, successive Directors-General of CERN, take G. Stoltenburg, German Minister for Scientific Research, through the CERN complex.

J. Van Offelen, Minister for Economic Affairs, and M. Toussaint, Minister of Education in Belgium, are shown new equipment being developed.

Italian Foreign Minister A. Fanfani, accompanied by members of the Italian delegation, study the control room of the PS.

President Franz Jonas of Austria is taken to see the experiments on the floor of the PS by W. Kummer, who combined the functions of professor of physics at Vienna University with those of chairman of the CERN finance committee—a role he filled admirably.

would be no significant net economic benefit in having the project in the UK (assuming full employment) and, on their alternative assumption that the project would adversely affect government stabilization policy, there would be a considerable net disadvantage. Moreover, they expressed the opinion that the spin-off benefits would be negligible and even if British firms did secure contracts for supplying equipment, the economic advantages would probably be small. They seemed to be studiously unaware of the UK's perennial balance of payments problem, or of the stimulating effect of international projects on national design and development at the industrial level, an influence that has more recently been quantified by a study undertaken at CERN by an Austrian physicist/economist. From data provided by the companies supplying equipment to CERN, he has shown that the economic 'utility' to these companies of doing work for the Organization averaged out at something like four times the value of the contracts that had been placed with them.

In spite of this negative comment on the value to the economy of the project coming to the UK, it was possibly a document published by CERN, making it clear it was highly unlikely that the UK site would be chosen, that turned the scales.

In December 1967, one of the very few really tactless documents produced in the huge stock of paperwork that was now amassing, was presented to the CERN council. Three delegates from countries not offering sites had been asked to prepare a comparative report on the nine sites that remained in the lists; Germany, having resolved its internal problems, put forward Drensteinfurt, near Munster, as its best offer, and the Northern Italian faction having gained over the Southern, the Italian offer was now firmly Doberdo near Trieste.

As information, the panel had the written replies to a standard questionnaire—replies which varied from the dry and brittle to the florid style of a publicity brochure. They attempted to grade the sites under three headings. Characteristics relating: (1) to the construction and development of the laboratory, (2) to the technical operation, (3) to the willingness of people to come and work there. Such a document was of course highly subjective and of major political significance, and soon after its publication it was quietly forgotten, but the damage had perhaps been done. The British offer of a site at Mundford, near Cambridge, not surprisingly got a 'gamma' on technical grounds. The member states had been asked to recommend a rock site in which a tunnel could be bored which would house the main accelerator, but the UK offered a site in chalk where the water table was at a depth of only 8 metres. Admittedly, the Americans were building a machine on a cut and fill method, but this was not proposed for the European machine. What was harder to stomach, was a 'gamma' for the third category, and comments such as

3/Site for Agreement

*The Project made steady if slow progress up to the fateful day in June 1968 when Sir Brian Flowers (**right**) with uncharacteristic gravity was obliged to announce the decision of the UK government to withdraw, but hoped personally it could continue. He is seen here with the French delegation of F. Perrin (**left**) and J. Martin.*

"it can be very depressing to live in a place where the sun seldom shines" were hardly likely to win friends in government circles. Only one site gained an 'alpha' in each of the first two categories: the French proposal of Le Luc, just north of St Tropez, and it was difficult not to feel that the CERN personnel's enthusiasm for the Côte d'Azur had been getting a little out of hand, and that they had been exerting undue pressure behind the scenes.

Whatever the ins-and-outs, it was Flowers's unhappy task to announce to Council in June 1968 that Her Majesty's Government had now decided in the light of its other commitments that expenditure on this very large project would not be justified. Then, to the astonishment of the assembled delegates, Flowers made a

Many were the private discussions held in the corridors of CERN to see how the Project could go on in the temporary (it was hoped) absence of the UK. Mervyn Hine (left), Adams's right-hand man on the PS and then director of CERN for planning, talks to the German delegation of H. H. Haunschild (centre) and W. Paul, professor at Bonn University and former head of the CERN Physics I Department.

personal statement on behalf of himself and Professor (now Sir Denys) Wilkinson (Chairman of the High Energy Physics Board of the SRC) that they, with the full support of the Science Research Council, hoped the project would still be pursued. The British nuclear physics community still considered that it should be given top priority even to the point of closing down a national accelerator, and would continue to urge their Government to participate at a later stage. The mood, instead of being one of despair, was transformed into one of euphoria as the delegates began to re-think how the programme could be modified so that it could start on a more modest footing and, when Britain joined later, be restored to its original grandeur.

3/Site for Agreement

For the president of Council, G. Funke, these were difficult times as he discusses with Gregory and Martin how the Project could be revised to go ahead with a reduced number of participating states without increasing their contributions. No delegation could go back to its government and start the long process associated with a demand for a new allocation of funds.

There was certainly, also, no less astonishment amongst the delegates when Brian Flowers a few months later, in spite of his personal stand against his government, became Sir Brian. This was generally interpreted amongst the other member states as being typical of British fair play, or of the British sense of humour.

In hindsight, it is doubly ironic that Sir Brian, one of the firmest devotees of CERN, should have been the instrument for precipitating it into its most dangerous phase of development, with the delegates unanimously embarking on a new adventure which would undoubtedly have destroyed the Organization if it had succeeded.

The new idea that formed was for a restricted number of the member states to go ahead with this great new enterprise on the assumption that others would join. Little heed was given to the divisive effect this would have on the attention paid to CERN, Geneva, and CERN, elsewhere, and the catastrophic effect on the Organization when budget economies began to bite.

Soon after the announcement of the British withdrawal, Germany and Italy agreed to come in, and they were followed towards the end of the year by Switzerland. The original Three reserved their position, but essentially, the project now became the project of the Six, downgraded to a cost of 1431 million Sw.Fr. at 1969 prices, and the design modified in such a way that the starting off energy would probably be 200 GeV, but the full programme could be restored at any time should the recalcitrant member states decide to come in.

A new stimulus to the urge to get on at any price had come from the decision of Congress in the United States to authorize the construction of a 200 GeV proton synchrotron at Batavia in the Middle West. The man who had been appointed to head this project, R. R. Wilson, had already shocked the conservative souls in the high energy physics world on both sides of the Atlantic by his proposals. In Europe, characteristically, his design innovations were treated with suspicion, the programme time-table as a piece of American advertising, and the whole project as in rather bad taste. The wiser elements were, however, aware of what American intelligence and industry once unleashed could do together, and if Europe dillied and dallied much longer it stood fair to build a machine of an energy range from which the exciting physics results had already been creamed.

The capital problem now became the choice of site, and a feverish round of new measurements on ground stability, water drainage, and the like were started at each of the five proposals now in the running. Switzerland, of course, was not offering a site, and although Austria made certain perfunctory efforts to promote its own, it realized early on that Göpfritz was number five. Belgium, on the other hand was extremely anxious that the site that they had chosen at Focant in the Ardennes should be selected, as it was the central pivot of a regional development scheme which had been hanging fire for many years. They believed that they had the support of France, who had been won over, it was said, in a series of tortuous negotiations which led to the purchase of Mirage fighters. The newspapers were full of the deal that had been concluded between the two countries, but it is doubtful whether there was any formal commitment. France was sitting pretty though. Le Luc had received an excellent report, there was still a sizable group of influential civil servants who saw themselves settling down on the Riviera, and it fitted nicely into their own southern development scheme. On the other hand should St. Tropez prove to be too French to be acceptable, they were equally happy with Focant, for which Belgium would have to provide all the infra-structure, but which would provide work for as many Frenchmen and French industries as if it

3/Site for Agreement

were on French soil. Germany with its proposal for Drensteinfurt, on the edges of the Ruhr, was grimly hanging on, getting more and more sullen as the lack of interest of its partners became more evident. In these circumstances, the Italians were hopeful that a deadlock would develop between Belgium and Germany, which would result in the only possible compromise: their own site near Trieste being chosen. The geological and geophysical examinations of the sites were costly and far-reaching. Shafts were dug, galleries hewn out, and these were solemnly visited by the members of Council and the SPC (Science Policy Committee). Ever more sensitive measurements were made of such things as ground movement, for as Sir Alec Merrison remarked drily to a puzzled journalist during one of these visitations, "If you have an essentially political problem on your

Members of the Committee of Council and the Scientific Policy Committee decide that they should see for themselves the five sites that were left in the running. At Doberdo, a study of the railway tunnels in the region seemed to be relevant and provided some light relief from political problems.

hands, but there is something you can measure, you go on measuring it with ever more precision; this does not, however, mean that it is important".

In the midst of this silly season, Council had taken the very important and wise decision to appoint a project director. They had been able to persuade John Adams to pick up the pieces of the project, and to see if he could put them together with a view to taking over and becoming Director-General of the new laboratory. Up to this time it had been Bernard Gregory who had had the impossible task of being both Director-General of CERN (Geneva) and co-ordinator of the new programme. A *Polytechnicien par excellence*, with a huge capacity for drive and work, and with a special gift for being able to pick out of a complex, disconnected discussion the essential elements, he found that the job was still too much for one person.

Abandoning then a secure position with the UK Atomic Energy Authority, Adams came to CERN again in April 1969, and set about the delicate task of bringing order into the political scene. On the technical side, he encouraged a new study to be made of the comparative costs of different approaches to the main magnet system, and set up a Machine Committee consisting of physicists from all the European accelerator laboratories to help in the work, incidentally ensuring the 'involvement' of all the laboratories participating.

By June, at the Council meeting the project was looking healthy. A programme for the construction and bringing into operation of the laboratory was approved, as well as a procedure for the selection of the site and for the establishment of the programme. There was by now a question mark over the French participation, but when confirmation of France's participation came at the beginning of November it looked as if all was over bar the shouting. There was still the site to be chosen of course, and lobbying was intense, but Council had after all agreed a system of voting which would lead to an explicit and, therefore, fair choice. All was set for the Council meeting in December.

But it was not to be. Germany, with a directness that shocked its partners, announced quite brutally before the December meeting that if the choice were not

Opposite
It proved impossible to find a compromise on the choice of site for the reduced Project, and Adams arranged for samples to be taken of the rock underlying the region alongside the existing laboratory. He then proposed that the new accelerator should be built there, which would enable the Project cost to be halved by making use of existing facilities and services.

3/Site for Agreement

Drensteinfurt, then Germany would not be participating, and being the largest contributor, this meant no project.

The announcement by Germany marked the end of an era—not of high energy physics, but of political standings in Europe. Germany had been experiencing her 'economic miracle', but her partners inside the Common Market, and her colleagues outside, made it perfectly clear that her role was to pay in proper proportion to her wealth, and then to keep quiet. Germany was regarded as one of the *nouveaux riches*, definitely not out of the top drawer, who should speak only when spoken to. Those in Germany concerned with the site selection felt they were being treated as 'second-class citizens' who should know better than to try and push their claims in the company of their 'betters'.

It is more than probable that the crudeness of the German announcement was designed to shock, although it was popular to interpret it as an example of stumbling diplomacy. If it was deliberate it certainly achieved its end, and from that moment German sensitivities and aspirations had to be treated seriously, and the German dossier returned to the top drawer. The feeling at CERN was one of real dismay, but it was hoped that a meeting of Ministers early in the following year might find some way out of the impasse. This meeting never took place, and the European 300 GeV programme seemed to have died, wasting seven years' painstaking work.

But on these foundations the project of the new Super PS was built, a project which was to reunite the member states of CERN, consolidate the high energy physics research of Europe, and give to CERN a homogeneous structure able to absorb the stresses imposed by the economic restrictions of a Europe in recession. The essential idea was simple. Instead of building a new laboratory somewhere else in Europe, it was to build a new machine alongside the existing CERN, use the existing synchrotron as the injector, existing experimental facilities for some of the experiments, and set out all the very high energy experiments along two lines, rather than prepare a whole series stretching away from the main ring. The new money needed from the participating States instead of being in the order of 2000 million Sw. Fr. would be something like 900 million spread over eight years.

A huge amount of work had gone into the evolution of the basic ideas, and the concept of a separate laboratory was deeply ingrained. Adams had to be extremely careful in introducing his own radical proposals. While the Machine Committee continued to work on the comparison of designs for an accelerator to be built on a separate site, Adams made discreet approaches to a number of governments, and especially Germany, to sound out their reactions. When these proved favourable, and the Scientific Policy Committee had also given its assent, a meeting of ECFA

3/Site for Agreement

was called to consider the idea. A stormy Sunday session followed, at which many delegates expressed dismay at the prospect of CERN (Geneva) being strengthened and the idea of the second laboratory being abandoned. It was only with difficulty that the meeting was persuaded to give a grudging acceptance to the proposal should no other route seem feasible. It was then left to Gregory to inform the CERN staff and parry the battery of questions that his announcement provoked. Adams had left for the US. By the time he returned the storm had calmed and the physics community had begun to rally round the one proposal they were recognizing might prove acceptable to all the member states. There were still, of course, individuals who complained they had not been considered or consulted, but it was not long before the ranks closed, and ECFA and high energy physicists were making a combined effort to push the project through. With good grace the Machine Committee now turned its attention to studying the implications on the design parameters, and in haste set about a new assessment.

Crucial to the new proposal was the reaction of the United Kingdom, and here Adams could count on the vigorous backing of the Science Research Council and the personal support and influence of Sir Brian Flowers. Most importantly, a new government had taken office and a change of mind could be made not only without loss of face, but almost as a matter of principle. Margaret Thatcher, then Secretary of State for Education and Science, came to visit CERN in September, and won many admirers for her charm and scientific grasp alike. She left behind the clear impression that Britain was back in pure science again.

There was still the problem of the site. A confidential sampling of rock underneath land adjoining CERN had indicated there was a reasonably homogeneous mound starting some 20 metres down, which would accommodate a machine of maximum diameter about 2·2 km. This land crossed the French/Swiss frontier, although it was largely in France. Two countries, therefore, would be involved in making this land available if they were willing, and within these countries there were hundreds of property owners who were going to be affected.

Paris was less than delighted. The new site was going to mean a considerable expense, a great deal of Ministerial effort, and the project would probably benefit the Swiss economy more than the French. However, with good grace, the French authorities bowed to the inevitable, and got down to seeing what could be made available and when. In Geneva, the authorities were relieved. They had been only too conscious of the depressing effects the construction of a new laboratory would have on the activities at CERN (Geneva), and had been making strenuous efforts to attract the European molecular biologists to come to CERN to take over the running in the years to come. Nevertheless, appropriating land in Switzerland

The attitude of the UK was still crucial and the turning point was probably a visit made to CERN in September 1970 by Margaret Thatcher, then Secretary of State for Education and Science, when each side succeeded in impressing the other. In the photograph (**from left to right**), *Jim Allaby explains to Mrs. Thatcher the intricacies of an experiment in the hall below; J. B. Adams, at that time Director-General designate of the Project; T. G. Pickavance, director of the Rutherford Laboratory and chairman of ECFA; W. K. Jentschke, Director-General of CERN; Sir Brian Flowers, chairman of the Science Research Council.*

is not a simple process; the different responsibilities of the Confederation and the Canton ensure there is always some hard bargaining to be done between the two, and the Canton may not always turn out to be the winner.

3/*Site for Agreement*

Finally, in February 1971 the member states of CERN agree to the Project documents and the delegates from all but Greece (and temporarily Denmark) were able to hand in their governments' assent to the proposals put forward for an 8-year programme of construction and operation, with a ceiling cost of 1150 million Sw. Fr. at 1971 prices. In the photograph the Netherlands delegate J. H. Bannier hands over his country's document to Mme. de Modzelewska of the CERN directorate.

Nevertheless, in spite of these difficulties, it seemed clear by the end of the year that both France and Switzerland would make the necessary land available at a nominal figure, and necessary site supplies of electricity and cooling water would also be laid on. The Big Four were ready now to support the project: France, Germany, and Italy had been rejoined by the UK, and all seemed set for the final decision to be taken in December 1970.

"And now, John, all you have to do is build it!"

But fate had one more trick to play. Britain had made it a condition that her participation by the member states of CERN in the new project should be massive, and crucial to the UK's acceptance was the agreement of the Netherlands and Sweden. Their agreement was by no means a certainty. Few countries in Europe were not experiencing serious economic problems, and a number were becoming alarmed at the proportion that high energy physics would take up of the budget they were prepared to allocate to pure research. Nevertheless, the Netherlands with some reluctance announced she would join, and was ready to sign the agreement. But Sweden was not, and it was only at the adjourned session of Council held on 19 February 1971, that finally ten member states of CERN agreed together on an eight-year programme for the construction of a super proton synchrotron and associated experimental facilities at a ceiling cost for the project of 1150 Sw. Fr. Strictly speaking, the figure was slightly less, as Denmark was not prepared to agree, and for some months remained outside the fold. By the end of the year, though, with the single exception of Greece, whose contribution to the organization is in any case artificially reduced by the other members to only 0·46 per cent, the family was now complete, and CERN and high energy physics in Europe were back on the right rails.

4/Enter the Builders

When the CERN Council finally gave its assent to the project, it approved in particular its duration and the financial envelope. The participating states bound themselves to stay with the programme until it was finished, having agreed the maximum overall cost and the yearly profile of expenditure, albeit with a little latitude. There was no carping about the odd per cent, and without discussion, a sensible amount was accepted for contingencies.

The least clearly defined element was the machine itself, as the Project had always been referred to as the 300 GeV Programme; but the White Book—or, more formally, MC 60—which brought together the conclusions of the studies made by the Machine Committee and its 14 Working Groups, included a number of options about ultimate energy, and the means by which this energy would be achieved. Happily, there were no formalists determined to dot all the i's at this stage. What was already clear, however, was that the machine energy was unlikely to correspond to the programme title, and only in passing would the figure of 300 GeV have much significance.

Nevertheless, certain of the machine's characteristics were prescribed. The existing complex of accelerators making up the CERN PS would accelerate some ten million million protons up to an initial energy of about 10 GeV. These would then be directed to the new machine and injected into a 2·2 kilometre diameter ring—the main ring of the SPS. Over a period of about one third of a second, the protons would be re-arranged into several thousand bunches spaced evenly round the ring, following which, acceleration could begin and energy added at the rate of over 100 GeV per second. Once up to the desired energy, the protons could be rapidly ejected or slowly extracted from the ring during a period of, say, half a second, after which the magnets would be de-energized and made ready to repeat the operation. A full cycle for an energy of 400 GeV of the SPS would take from six to eight seconds. As the time needed for the pre-acceleration in the PS would be of the order of only one second, the PS could be used for experiments at lower energies while waiting for the SPS to complete its cycle.

Europe's Giant Accelerator

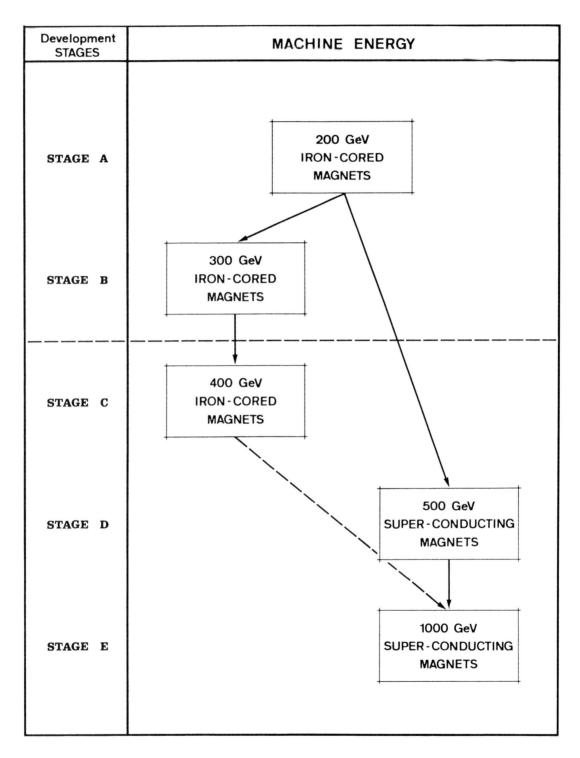

4/Enter the Builders

Certain materials at very low temperatures become superconducting and offer zero resistance to the passage of an electric current. A high magnetic field destroys this property and until fine filaments of superconducting material were produced it was only possible to make d.c. magnets based on the principle. (Photo Rutherford Laboratory)

Opposite

Although the financial envelope of the Project was completely explicit, a number of possible schemes were put forward based on the construction of a 200 GeV machine first, followed by developments which would be determined by the technical progress made, particularly in the field of superconducting magnets. These, for a given diameter of machine, would allow higher energies to be achieved than was possible with iron-cored magnets. In the event, none of these routes was selected, and a 400 GeV machine was built in one step.

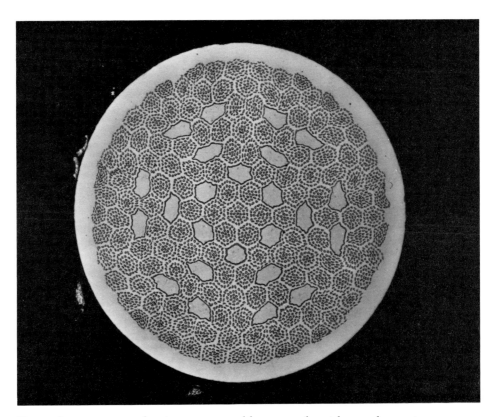

To make a superconducting magnet able to work with an alternating current, filaments of wire are twisted together into fine cables and these cables are embedded in a copper matrix which acts as a heat sink, and as alternative conducting path if, locally, a filament should lose its superconducting properties. In spite of the successful scientific development of a.c. superconducting magnets, it was not considered that they were sufficiently developed industrially to justify their use in the main ring of the SPS.

Situating it next to the existing laboratory of CERN offered a number of advantages, in eliminating the need for an injection accelerator and exploiting services already in being, but it imposed serious limitations on certain aspects of the design. A further complication was the tempting lure of superconductivity. CERN had built the very first strong-focusing machine using iron-cored magnets, and was anxious not to be the Organization to build the **last** obsolescent member of the series.

4/Enter the Builders

For many years, it has been known that certain materials when very cold present zero resistance to the passage of an electric current. Consequently, huge currents can be carried by quite small wires without their heating up and melting. If, however, the conductor loses its super properties even for a moment, it promptly melts, so that the practical safe application of the effect is far from straightforward.

The strength of magnetic field which can be produced between the pole faces of an ordinary iron-cored electromagnet is limited by the nature of the iron itself. The iron concentrates the field and magnifies the effect that would be produced by the activating coils alone, but only up to a strength which is fixed by its detailed composition. If we wish to go beyond this saturation value, then the iron has to be dispensed with and we lose the amplifying effect. To offset this, we must have more turns in the coils, but this requires space and a law of diminishing returns applies; alternatively, we must increase the energizing current, which requires thicker conductors to prevent their overheating, and we are almost back where we started.

Superconductivity offered, in principle, one way of increasing the strength of the guiding magnet fields in a synchrotron, so raising the maximum energy that could be attained in a ring of a given diameter. Also, it could offer savings in operating costs, because of low electricity consumption. Unfortunately, superconducting materials return to the normal state if the magnetic field to which they are subjected goes too high, and it is in the nature of things that those materials which can cope with the highest fields are the most difficult to work. Nevertheless, magnets had been built which produced magnetic fields 5–6 times stronger than could be obtained by other means. But these were d.c. magnets which took a long time to reach a steady state; any attempt to change rapidly the current in them resulted in the coils ceasing to be superconductive. However, research had shown that if the conducting wires were drawn extremely fine and cables were made up of many wires twisted round each other, all embedded in, say, a copper matrix, quite rapid changes could be made in the current passing, without the conductors reverting. Techniques of construction were also being developed that would take care of the huge mechanical forces generated in the delicate windings by the field, which seeks always to make the coil collapse on itself.

It appeared prudent to leave open the option of installing superconducting magnets in the main ring of the accelerator for as long as possible, in order to be able to arrive at the maximum energy technically feasible. There was also an inherent attraction in the possibility of building the very first big example of a brand-new technology.

*The CERN laboratory in 1971 looking across the terrain under which the SPS would be built towards the Jura mountains. A little left of centre is the 200 m ring of the CERN PS and, to the left, the mound covering the 300 m ring of the ISR. The Franco-Swiss border follows roughly the line of the woods which run into the laboratory site and then follows the CERN fence (**lower left**) to the corner. France is upper and left; Switzerland is lower and right.*

So Council accepted a design report which contained at least five strategies, and left it to the newly created Project team to furnish firm engineering proposals within a year, deferring the decision on superconductivity until the last possible date—1973. In essence, the schemes put forward assumed that to begin with a synchrotron of 200 GeV would be built, which could be used for experiments located in the existing CERN complex, while preparations were made for raising the energy to some higher figure and the new experimental area was being built.

The ruse adopted to pause at 200 GeV without restricting later possibilities, was to fill only half the available space in the main ring with conventional magnets, leaving clear gaps between the individual units. Those magnets installed would be

4/*Enter the Builders*

Adams in front of a photograph of the Geneva basin with the ring of the SPS traced out and showing the injection line from the PS, the ejection line to the existing buildings (**bottom right**) *and* (**lower centre**) *the beginning of the beam line to the north area.*

Europe's Giant Accelerator

placed symmetrically round the ring, and subsequent additions, in their turn, would be put in symmetrically. Within the original budget proposal, half the remaining space could be filled by conventional magnets to raise the energy to 300 GeV, and, if more money was forthcoming, the remaining quarter could be filled to bring the maximum up to 400 GeV. If, on the other hand, it was believed that superconducting techniques had advanced sufficiently, then, after 200 GeV, the spaces could be filled with these high field magnets which, operating alone, would give 500 GeV. If in addition, the iron-cored magnets were then replaced, 1000 GeV was within reach. This was a magic figure which inspired the physicists and frightened the delegates, who knew well that every increase in primary energy brought a corresponding increase in the cost of doing research.

With all the complicated calculations that go into the design of accelerators, this was the first time that it had been appreciated that doubling or halving the total length of bending elements in a ring which is measured in kilometres, where the individual elements are measured in metres, changes the orbit by a few centimetres only. Within the limits set by the mean ring diameter, a wide variety of configurations could be accommodated. The missing magnet concept, as it was labelled, allowed all these several ways to be left open for the time being, and it was typical of CERN that the Council should accept, when the time came, without demur, or even debate, a strategy altogether different from any of those included in the original document.

With so little of the ultimate potential of the machine determined, the question of a name posed some problems. It could hardly be referred to by the programme title, or by a top energy not yet chosen. The press had branded the complete project long ago as Super CERN, which had caused a lot of dismay amongst the more sensitive souls in the existing CERN, as this seemed to imply that the first establishment was a little less than super. Nevertheless, the biggest machine in the world could hardly come into being without some sort of appellation, so the initials SPS came into use to signify Super Proton Synchrotron, Superconducting Proton Synchrotron (if that should turn out to be the way of development) or, in the French language, Synchrotron à Protons Souterrain, emphasizing that the machine was to be constructed in a tunnel out of sight of local people.

MC 60 was a remarkable document in view of the fact that it summarized the work of over 100 physicists scattered across Europe, who, in their turn, had mobilized the help of countless other experts in the universities and national

Opposite
Internal organization of the SPS.

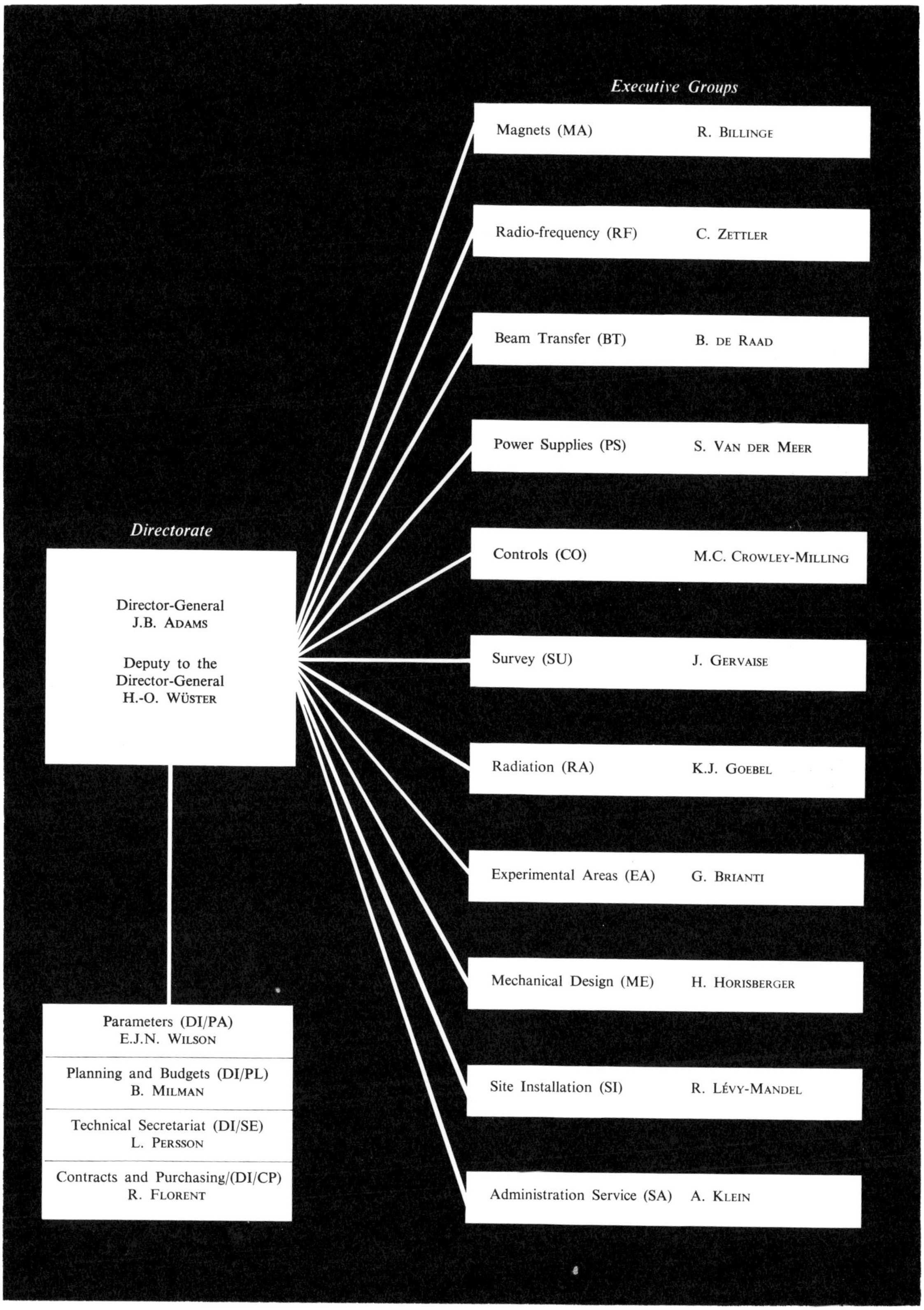

Directorate

Director-General
J.B. ADAMS

Deputy to the
Director-General
H.-O. WÜSTER

Parameters (DI/PA)
E.J.N. WILSON

Planning and Budgets (DI/PL)
B. MILMAN

Technical Secretariat (DI/SE)
L. PERSSON

Contracts and Purchasing/(DI/CP)
R. FLORENT

Executive Groups

Magnets (MA)	R. BILLINGE
Radio-frequency (RF)	C. ZETTLER
Beam Transfer (BT)	B. DE RAAD
Power Supplies (PS)	S. VAN DER MEER
Controls (CO)	M.C. CROWLEY-MILLING
Survey (SU)	J. GERVAISE
Radiation (RA)	K.J. GOEBEL
Experimental Areas (EA)	G. BRIANTI
Mechanical Design (ME)	H. HORISBERGER
Site Installation (SI)	R. LÉVY-MANDEL
Administration Service (SA)	A. KLEIN

research laboratories. The design was to change in many details, but the essentials were established, and, as time went on, it proved to contain few of the monstrosities normally associated with the products of committees. The reason may be that although the Machine Committee was big, and its members were widely spread, the number of people engaged full time on the Project in 1970 was just four, plus those tied up with the geological assessment of the site.

Apart from Adams, Director-General designate of the Project, and his bright, multi-lingual Dutch secretary, Frieda Dockheer, known to all CERN as Pepi, a woman with a mind of her own and a mission to keep the others in their place, there were two physicists to hold the strings to the various Working Groups. One was Ted Wilson, a brilliant, dedicated physicist with a husky voice and dry, penetrating sense of humour. He had come to CERN from the Rutherford Laboratory in 1967, abandoning high energy experiments to work on the injector in the original design concept. Unhappy with the stagnation in the main accelerator design, now beginning to look old-fashioned, he had begun to work out the main features of a more streamlined version, but received little encouragement from the 'old guard' who were anxious not to give the impression that any improvement on the 1964 ideas was needed. Adams, when he came, was more receptive, and Wilson was pressed to push his calculations and cost estimates further. As a result, the Machine Committee was able to consider an alternative lattice (as the array of magnets in the main ring is termed) to the one drawn up in the very early days of the Project. When Adams decided that the SPS could only be built at CERN (Geneva), it was Wilson who carried out in great secrecy the initial lattice calculations. The other physicist who was with the Project from the beginning was Clemens Zettler, a radio-frequency wizard from Munich, with a passion for large American cars. He had come to CERN in 1964 from Radio Free Europe, where he had been responsible for the technical performance and development of transmitters and aerials. At CERN, he worked steadily on the accelerating system for the new machine, going on quietly with his experiments throughout the tumult of the various Project upheavals, unruffled, deliberately courteous, calm and balanced, whatever was going on.

This was the minute team at the centre of the outline design. But with the Project agreed, the time had come for the detailed work to begin, parameters to be frozen, drawings prepared, offers from industry requested, and so on. The Machine Committee remained in existence as an advisory body, and later carried out a thorough theoretical analysis of the behaviour to be expected of the accelerated beam once the detailed design was well advanced, but from here on the Project team proper had to be assembled, and the site acquired.

4/Enter the Builders

Hans-Otto Wüster, Deputy to the Director-General of the SPS.

At a Monday morning management board meeting the work of the various groups was coordinated. Round the table from right to left are R. Billinge, magnets; C. Zettler, radio-frequency; H. Horisberger, mechanical design; (hidden) B. Milman, planning and budgets secretariat; G. Brianti, experimental areas; R. Lévy-Mandel, site installation; J. Gervaise, survey; A. Klein, administrative service; S. Van der Meer, power supplies; K. J. Göbel, radiation; B. de Raad, beam transfer.

Over the previous months, discreet approaches had been made to experts in accelerator building across Europe, and the majority of the eventual Group Leaders selected in consultation with the leaders of the European physics community. The new staff had now to prepare to move to Geneva, sort out their family affairs, their children's education—always a thorny problem, even for the French-speakers (*francophones* in French) if they had not previously been resident in the area—decide what to do about a house, and whether to sell up at home, with the knowledge that once the job was finished there would be no automatic continuation.

4/*Enter the Builders*

From the bottom of the management board table. To the left of Adams in the photograph is N. Blackburne, personnel, and to the right of Wüster, partly hidden, L. Persson.

Immediately under the Director-General was a deputy, and under them eleven Executive Group Leaders, with for the most part responsibilities reflecting the different components of the machine, rather than the techniques involved. For example, magnets abound not only in the main ring, but also in the systems used to get the beam from the injector into the ring, and to get the accelerated beam out. Similarly, the lines leading to the experimental areas, as well as the areas themselves, bristle with magnets, but it was thought better to divide duties by machine function and not style of equipment, relying on a regular Monday morning board meeting to iron out the interconnected problems and promote standardization.

An early choice was the deputy to the Director-General, Hans-Otto Wüster, a heavy-weight in all respects, which the ensuing years of dieting and long hours have failed to obscure. Greatly respected on the directorial board of the Deutsches Elektronen Synchroton (DESY) in Hamburg from which he came, he brought to

André Klein, head of the administrative service with special responsibilities for maintaining good relations with the local authorities and Paris.

P. Tröndle, who spent much of the early period discussing with landlords the exact location of the international territory and how it was going to affect their particular property. In spite of the number of landowners touched by the Project, and the speed with which the land was acquired, relations between CERN and its neighbours remained remarkably good.

the Project a solid experience in laboratory management, a first-class scientific brain, a profound knowledge of and regard for computers, and a reputation for being jovial in private and breaking tables at meetings where the discussions seemed protracted. Straight in all his dealings, passionately dedicated to the Project and to the CERN ideal, he quickly won the same respect at CERN as he had had at DESY from staff and delegates alike and, to everyone's surprise, broke very few tables in the process. It was he who had swayed the stormy ECFA meeting when the Geneva site for the SPS was first proposed to it, and formulated the resolution which marked acceptance.

Crucial also to the project, especially the rapid acquisition of the site, and the maintenance of good relations with the host states, was the appointment of André Klein to the position of leader of the Site Administration Group. Formerly the sous-préfet of Gex, the surrounding area in France, he brought the skills of a training in the élite French administration, a detailed knowledge of French procedure, and an evident sympathy with the fears and aspirations of the local mayors and the landowners, shortly to be disturbed. His feeling for the sensitivity of the French dignitaries in the region (who referred to him as "our man in CERN") was also of major importance in clearing the lines to Paris. He was joined by Peter Tröndle, a bucolic Swiss whose untiring rounds of the local farmers did much to smooth the troubled ground (even if they played havoc with his own waistline), and Georges Stassinakis, a legal expert from Greece, who always looked indecently cheerful for a lawyer engaged on long drawn out negotiations of leases and statutes. Each over six feet in height, Klein, Tröndle and Stassinakis made a formidable trio when out together, quite eclipsing the neat athletic figure of Noel Le Vasseur, recruited from the French customs service, to handle the complex problems of shipping duty-free equipment from industry to the site, and subsequently moving it around the CERN area.

CERN already straddled the Franco-Swiss frontier, and had concluded special agreements with both States. But up to that time, the whole site was enclosed by one fence, and entry for personnel was from Switzerland, even if provision was made for goods arriving from the French side to enter CERN directly. Now CERN was to spread out over a territory some 5 km by 3 km, only a small part of which would be occupied by buildings. Enclosing the whole area would have totally disrupted the communications in the locality, and posed CERN a land management problem of formidable proportions. It was environmentally desirable, socially preferable, and technically advantageous that the majority of the land should be farmed as before.

The solution was to acquire the land immediately above and around the main ring and beam lines, apply building restrictions on nearby land to ensure no subsequent deformation of the underlying rock, and only fence in those islands needed to accommodate surface buildings. The implications for staff communications were evident. The internationally-controlled area was no longer continuous, and people must cross a national border to move from one part of CERN to another. No precedents existed, and the host states would have been within their rights to insist on the formalities being observed on every occasion. Instead, a tunnel was dug under the road linking St. Genis with the Swiss border, connecting open French territory with the CERN-enclosed site entered from Switzerland. The border gate was manned by CERN personnel. For the individual, the difference was not so great, except at times of peak traffic, but for the transfer of equipment the gain was immense.

Another problem of communication arose in relation to the telephone. Officially, all calls between the two laboratories would pass through an international exchange, although there is now direct dialling between Geneva and the Pays de Gex. Once more, the host states preferred pragmatism to protocol, and the new buildings were connected up to the old (Swiss) exchange, with a few additional lines linked directly to the French network.

4/Enter the Builders

To simplify the movement of personnel and, more especially, equipment between the laboratories, a tunnel was dug under the Geneva–St. Genis road. A single check point (**left**), policed by CERN personnel, controls the traffic between the old laboratory entered from Switzerland and the exit on to a public road in France which crosses the SPS ring. This exit is close to the entry of the civil engineering area of the PGC (**upper right**) now returned to arable land.

Life became yet more complicated in April 1976 when the French decided to introduce summer time, adding one hour to the Continental time adopted up till then throughout the year in both France and Switzerland. CERN as a result became a centre which not only crossed a frontier, but a time zone, too. Democratically, the staff was consulted, and by majority vote, while the clocks remained Swiss over the entire centre, the official hours of work were advanced half an hour.

Attached to the directorate were four secretariats. Ted Wilson looked after the machine parameters, and kept tabs on all small design changes to make certain that it was a complete machine, and not a series of isolated components; Boris Milman, a quiet, ascetic Frenchman, looked after the planning and budgets. This was to be no hit or miss affair, either in the honouring of budget commitments, or in the programming of the construction schedule. The components of the machine, once specified or explicitly defined, for the most part would be supplied by European industry, and subject to all the vicissitudes of a worsening economic situation. It was well understood that even when the equipment required was not of novel design or demanding novel techniques, the very size or quantity and consistency in precision needed would stretch industry to the limit. With so much that was innovatory to be made, there were certain to be slippages, delays in the supply of components and materials, and the programme would need continuous adjustment. A computer-based PERT program was developed to up-date continuously the design, ordering, checking, and installation schedule for a project that could not be fully defined until three years after it had begun.

CERN already had a purchasing department, but the load which the construction of the SPS was going to add could not be accommodated within the existing framework, and a special secretariat was needed. To head it came Roger Florent, of France, who had coped with the technical and administrative complexities of building BEBC, the world's largest liquid hydrogen chamber, at CERN, under a tripartite agreement with the Comissariat à l'Energie atomique of France and the Federal German Government. Although shy and self-effacing, and only really happy when battling out a force 10 gale in a small boat, Florent's cool attention to detail and quiet persistence were to stand him in good stead in the stormy waters of handling contracts with firms from more than a dozen nations.

There was also a technical secretariat, under Lars Persson, a slight, dogged, intensely loyal Swede, whose job was to publish the progress bulletin circulated to members of the staff, ensure liaison with the existing laboratory services and pick up the bits that fell to no-one else. He generally, with an assumed imperviousness, took the blame for any arrangements that went wrong, whether they concerned him or not.

4/Enter the Builders

In the early days, recruitment of suitable personnel was a first priority. Once on site they had careers to be followed, personal difficulties to surmount, and the practices of staff advancement in the various groups to be harmonized. Into this secretariat came Norman Blackburne, a merry, comfortable Irishman, never short of a funny story to match the occasion, or a firm word of caution for someone tempted to step too far out of line.

This, then, was the administrative framework within which the eleven Executive Groups were to operate. Nursing them all was Yvonne Henry, looking after the general secretariat, cheerful and pragmatic, for the first time since coming to CERN able to give full reign to her natural organizational talents. Surprised at nothing, the white laboratory coat she habitually wore suggested the consulting room of a clinic, rather than the local administrative centre; perhaps this at times seemed more appropriate.

Two Executive Groups, the Mechanical Design and Radiation Groups, had functions which concerned in varying degrees all that went on in the other technical groups. Heading the former was Hans Horisberger, a Swiss engineer of serious mien and cautious disposition. He had first joined CERN in 1954, on returning from a four-year stint in the USA, to work on the building of the PS. When that was completed, he left to join a printing machinery company, but returned in 1966 to lead the General Engineering Group of the ISR. His wide design and manufacturing experience fitted in well with the varied tasks that fell to him. In addition to the drawing office and workshop, he looked after the group responsible for the vacuum chamber in which the protons circulate, and the pumps which maintain the pressure in the system at one ten-thousand-millionth of an atmosphere. Later, he was responsible also for coordinating the installation work in the tunnels, and organizing the teams working down below.

The Radiation Group was headed by Klaus Göbel, from Germany, who on leaving the University of Freiburg came to CERN as the leader of an experimental team working on the synchro-cyclotron. In 1962, he began to specialize in radiation problems, learning not only the science of the subject, but also the practical implications of satisfying the evolving radiation codes of two countries, drawn up essentially to cover nuclear power and radioisotope applications, and not beams of very high energy particles. Trilingual (as is Horisberger), charming with an almost old-world politeness for one looking so young, he held firm views on the proper interpretation of international recommendations on radiation protection, and the way in which these should be implemented.

The radiation problems of the new machine were quite different from any that had been encountered before, at least in energy if not in kind. No tables existed

A special computer-operated PERT program was developed to follow the evolution of the project and simplify critical path analysis. The schedule was continually up-dated and checked through at the Executive Board meetings so that slippages were immediately noted and correction measures applied.

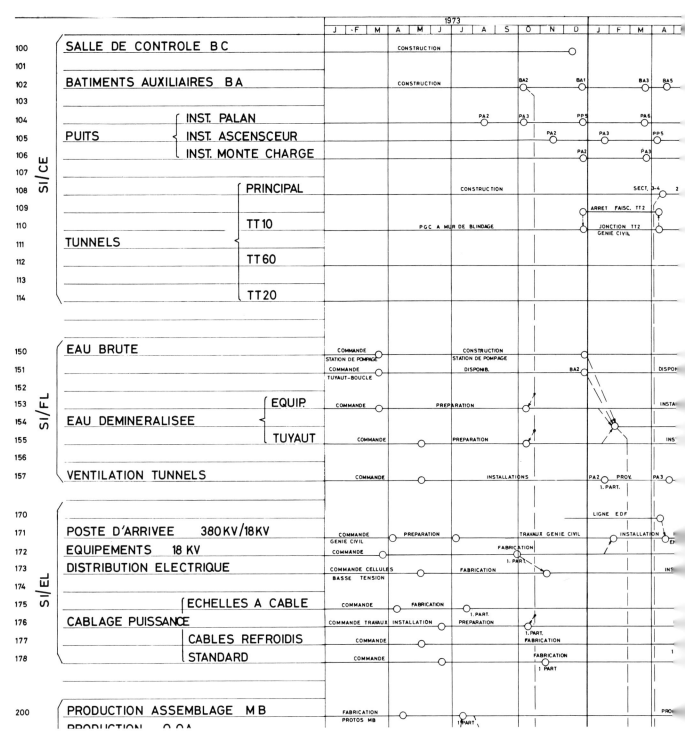

4/*Enter the Builders*

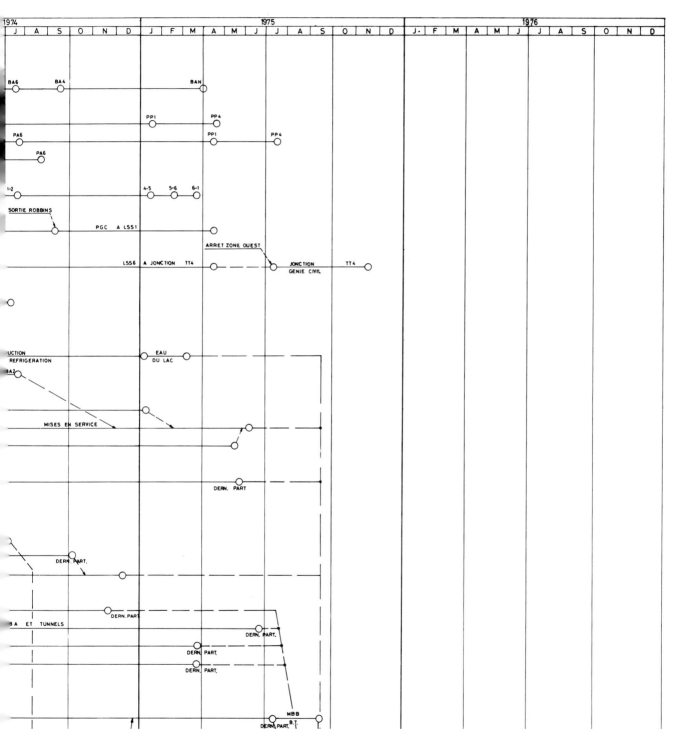

Hans Horisberger in charge of the mechanized design group which included the design office and workshops concerned with development as well as fabrication; the section responsible for the vacuum chamber in which the protons circulate and the vacuum pumps which pump it down to a low pressure; and the section concerned with installation which was required to work to a strict schedule, particularly in the tunnels where access was restricted and where movement of large pieces of equipment must follow a precisely pre-determined plan. Later he was to succeed Lévy-Mandel in charge of site work.

4/*Enter the Builders*

Klaus Göbel, in charge of the radiation group, had to ensure that no aspect of the SPS operation would present any hazard to either the local population or members of the CERN staff. He also advised on special materials for the areas subject to high radiation and set up the routine monitoring system around the site, beginning systematic measurements before the SPS became operational.

from which the effects of the high energy beams could be read off, and it was necessary to analyse theoretically the multiple event pattern of stopping a high energy particle, and to extrapolate data amassed at lower energies. The gross problem of protecting the public from the radiation produced by the accelerator was much simplified by the decision to place the machine deep underground, under far more rock and soil than would be needed to reduce the radiation to an acceptable level. Even in the experimental areas the beams could be directed, especially in the north area, either into the ground, or, in the case of secondary beam rejects, into the sky. There was still the question of activation of the air and drainage water, but above ground the need to isolate one experiment from another often imposed the most severe shielding criteria. Down in the tunnel, particularly in those areas where significant beam loss was to be expected, as in the injection or ejection regions, or in the target zones, the problem was to protect working equipment, and make it possible for people to undertake maintenance when the machine was shut down.

Consideration had to be given to the installation of remote-handling equipment, and to the shielding of sensitive apparatus, or in exposed points the use of special radiation-resistant materials, particularly for electrical insulation. All groups had to be aware of radiation aspects, and Göbel and his group were there to give expert advice, and set up the routine monitoring.

Although so many options as to final energy had been left open, none of the group leaders was in any doubt that he was personally responsible for the component design and budget control in his own sector, under the ever-watchful eyes of Adams and Wüster. The first job was to produce a definite design to be presented to Council at the end of 1971, leaving enough time for collation, translation and printing. It was, in effect, an up-dated White Book, but this time written by the people who would have the responsibility of putting into practice what was set down. The resemblance the final report, known everywhere as CERN/1050, bore to its forerunner was a fitting testimony to the quality of the studies of the Steering Committee and Working Groups, and the little band who coordinated their activities.

The decision on superconductivity, when it was made in the middle of 1973, was essentially industry-conditioned. Research on superconducting technology as applied to accelerators was mainly concentrated in three national laboratories—Saclay in France, Karlsruhe in Germany, and the Rutherford Laboratory in England. This was coordinated by the GESSS committee (Group, European, for Superconducting Synchrotron (now Systems) Studies), a collaboration that merited a much more laudatory acronym. After intensive study, the recommendation was not to consider superconducting magnets for this time. Scientifically, enough was

4/Enter the Builders

The beginning of the accelerating system, which involves five separate accelerators, is a 750 kV high voltage set which applies this tension between the hollow ball (**upper centre**) and the cage of the room which is earthed. In the square box is the source of hydrogen from which the protons are extracted and the auxiliary power supplies and controls which ionize the gas and strip it of its electrons. From this high tension set, the protons go through a linear accelerator (linac), a booster and the PS before injection into the SPS.

Layout of the SPS ring showing the linac, booster (PSB), PS and the transfer lines to the SPS passing under the PGC. The six auxiliary buildings house power supplies, heat exchangers and the local control equipment. Access to the underground ring is via lift shafts from these buildings. The beam lines to the West experimental area are shown (**upper right**) *passing beside the ISR, while the North*

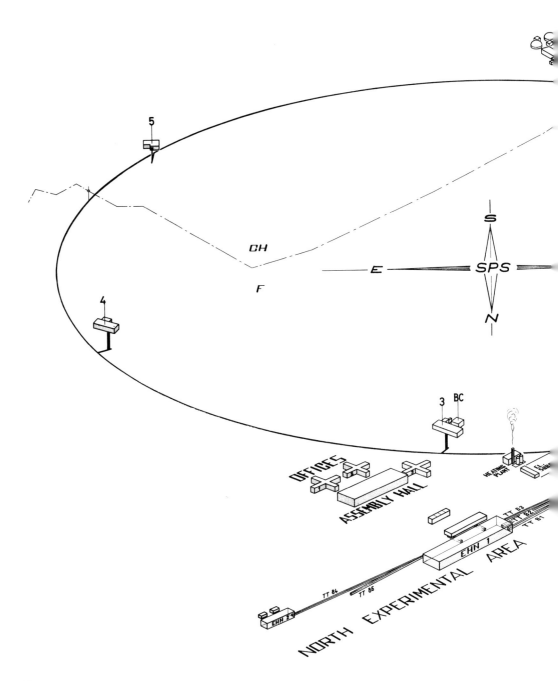

experimental area is located near the design offices and assembly hall. Power is brought in via a special 380 kV grid line to a sub-station in the same area close to the local heating plant. The control room (BC) is situated alongside auxiliary building 3.

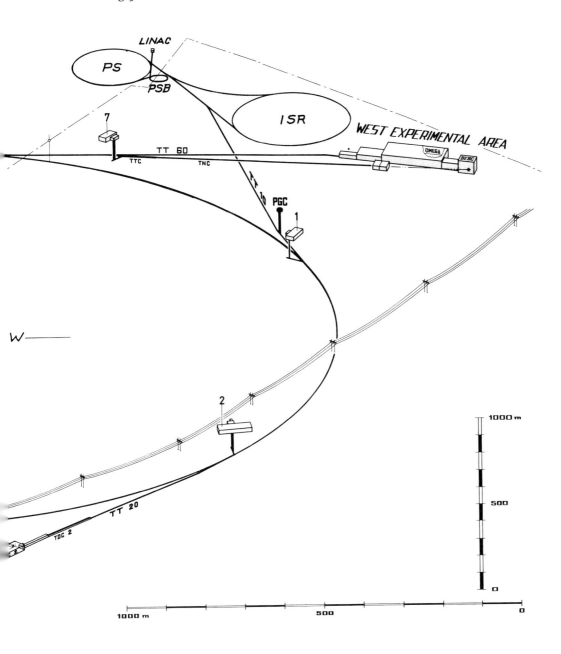

known to allow a superconducting synchrotron to be built with confidence, but the industrial problems of manufacturing on a large scale with the reliability and reproducibility necessary were far from resolved.

Instead, Adams proposed to Council to go straight to 400 GeV by filling the main ring with iron-cored magnets in one step, which meant a slightly later start on the research programme, but entailed no subsequent long shut-downs for modifications to the machine. He was ready to do this out of the funds already agreed. This was approved by Council, much relieved not to be drawn into a further technological gamble, at a time when funds all round were uncomfortably tight. A certain delicacy was needed to begin with, as the Dutch parliament had ratified the Netherlands' participation in the project in terms that rather explicitly limited the operational energy to 300 GeV, but this was sorted out, and the SPS could eventually be openly described as a 400 GeV proton synchrotron.

5/Down to Earth

While the administrators and the jurists were getting down to the problems associated with the acquisition of the land, the engineers had to define completely, explicitly, and exactly what the machine (the SPS) would look like and where it would be. The mean diameter of the main ring had been fixed at the maximum the site would take. This had to be, also, a simple multiple of the CERN PS diameter that was to act as injector. It was no good picking an arbitrary size and then finding that when protons were injected from the PS they were all squashed up together and great lengths of the ring were left empty.

There are two methods of taking protons from an accelerator: either extracting one bunch at a time, or peeling them off successively. The former method was well developed, and very high efficiencies could be obtained. For greatest convenience, the bunches would be whipped out, one bunch at a time and fed into the SPS, to finish up with a perfectly regular spacing. This is a smooth operation if the ratio of the diameters of the SPS and PS is 9 or 11. Either way, with the right steady rhythm, single bunches ejected from the PS on successive turns will space themselves evenly round the SPS. The physicists wanted the biggest ring possible, so the mean diameter was fixed at 11 times the PS diameter (11×200 m), 2·2 km.

This would require a tunnel nearly 7 km round, of a complex shape, because although a synchrotron is very nearly circular, a number of long straight sections are needed for the equipment which receives the protons when they arrive, which accelerates them as they orbit round, and which ushers them out to the experimental areas when the time comes. A certain regularity in the distribution of these sections is also desirable to avoid upsetting the circulating beam, and it was decided to divide the ring into sextants, providing one straight section for injection (No. 1), one for acceleration (No. 3), two for ejection to experimental areas (Nos. 2 and 6), leaving one for dumping the beam out of the way when not wanted (No. 4), and one spare for special instrumentation or anything else that might turn up (No. 5).

The ring was to be connected by tunnels to the PS and to the experimental halls, making about 10 km of underground work in all. The existence of the PS and the West experimental zone placed severe restrictions on the location of the SPS ring.

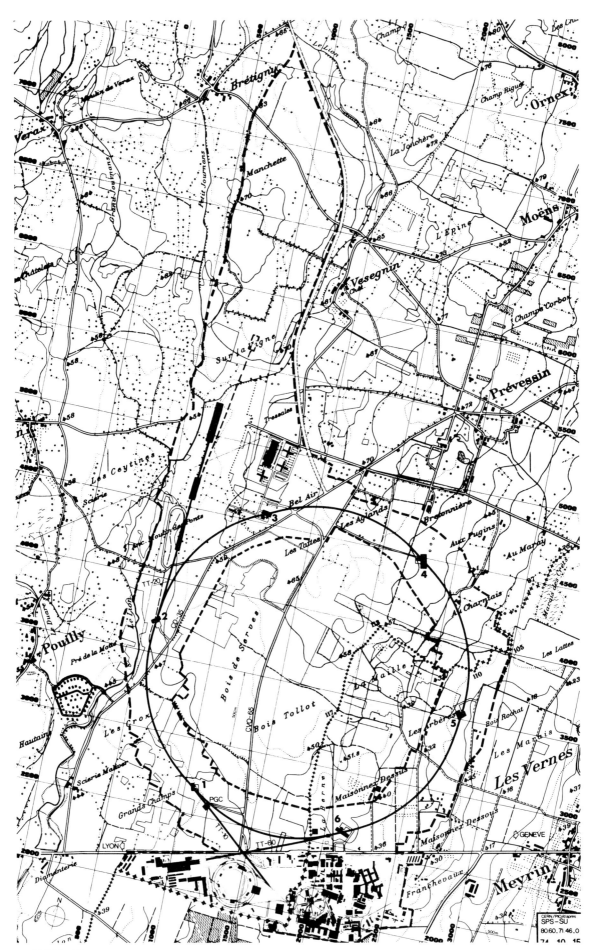

5/Down to Earth

In all other comparable installations it had been possible to establish the position of the principal element with a fair degree of liberty and to allow the dependent elements to flow from it. With the SPS, two independent fixed points had to be taken into consideration, and in the case of the PS with a precision measured in millimetres for the civil engineering, and in tenths of a millimetre for the placing of the beam line which would couple the two accelerators together.

The problem was not entirely new to CERN as the double rings of the ISR were coupled to the PS, and the specification for the placing of the ring components had been similarly exacting. The ISR, however, were only 300 m in diameter and less than half a kilometre away; moreover, they were built in an open trench which gave fair latitude to the builders and visible access to the surveyors. Nevertheless, it had been necessary to survey the site with great precision, and a system of monuments supported on good rock had been established and their positions relative to each other and to the CERN machines measured with great care.

For work of this type, there was already on site the necessary experience of high-precision large scale surveying in the person of Jean Gervaise, a tall, articulate Frenchman with a fondness for pontifical quotations, whose energetic activities in local government affairs had deepened his knowledge of the surrounding region. Adding to his basic qualifications of surveyor and geometer the academic and research traditions of CERN, with its enthusiasm for computers and the most up-to-date instrumentation, he had created there one of the most skilled survey-measurement teams in Europe with a reputation going far outside high energy physics. He had been heavily involved in the site selection studies, and personally had been responsible for the core sampling during the exploratory phase. Now, as head of the metrology group, he had the job of completing this underground testing and carrying out a precision geometric survey on the surface from which all other measurements would lead.

Opposite

The surface area required for the project formed an irregular figure 6, a little bigger across than the length of the existing laboratory, with the stalk stretching away to the north. Altogether some 412 ha were put at the disposal of CERN by France and 68 ha by Switzerland. Only a fraction of the site is fenced in, but because of the extreme stability required for the machine foundations it was necessary to have control over a band of land around the ring and a further tract to the north, to allow for future extensions to the beam lines there. In addition, building restrictions have been applied to the ring centre and other areas, totalling 509 ha in France and 63 ha in Switzerland.

Although to the casual observer the site looks fairly flat, there is, in fact, a change of level of some 50 metres. The core borings showed the underlying rock to be in the form of a smooth inclined ridge running NE–SW, the summit level dropping by about 25 m over the diameter of the ring. The rock is typical of the region between the Jura and the Alps, a grey molasse (incidentally, a name derived from the same roots as molar, and suggesting the grinding processes which went into its creation). It is a sort of sandstone mixed with clay, quite soft and crumbly, but in the mass it constitutes a geotechnical block of good stability.

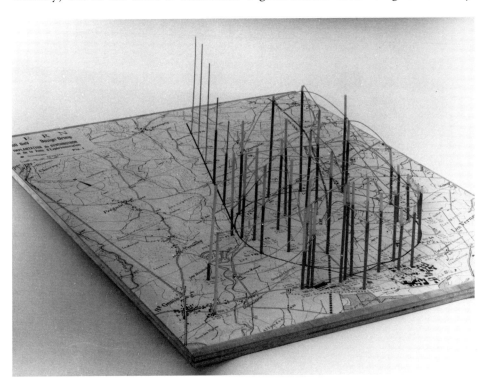

*The main ring was built in a horizontal plane relative to the vertical at the ring centre at an average depth of 40 m. The topography of the land above is irregular, the upper layers consisting of a varying depth of moraine (**shown in grey**) overlying the relatively homogeneous molasse rock beneath (**black**). To the west and east of the ring are valleys in the rock which limited the maximum ring diameter that could be accommodated on the site at a depth reasonably accessible from the surface, in particular for the injection line from the PS and the ejection line towards the west experimental zone.*

5/Down to Earth

Robert Lévy-Mandel, a general engineer and accelerator man by training, who was in charge of the whole of the civil works until he was made a director of the combined laboratories in 1976. Here he is explaining with characteristic vigour the construction work in progress in the north area to a group of delegates from the CERN Finance Committee.

Although cracked and fissured on the small scale, there is no water-table inside the molasse and it is covered by an impermeable overburden. As a consequence, it was safe to anticipate that tunnelling would be a dry operation, and the excavated part would not be exposed to the deleterious action of standing or running water which would cause the clay component to swell and deform as its water content changed.

Against these satisfactory qualities of the rock had to be put its sensitivity to mechanical loading. Elsewhere in Switzerland, in tunnels dug in similar rock the floor had been known to balloon upwards by as much as 20 cm as the strains in the rock mass adjusted to the reduction in the load. This had to be prevented in the foundations for a machine that demanded a stability of a tenth of a millimetre over 2 km. The tunnel would have to be lined with a strong circular vault, as soon as possible after the removal of the spoil, to re-establish equilibrium.

Over the molasse is a moraine of variable composition, and of depth varying from over 20 m to practically nothing. It contains, under the rich topsoil of the Rhône valley, a range of sands, gravels and clay with little water courses and scattered, isolated water-tables. This is messy and unreliable for digging, and those tunnels which had to pass through it came to be the most disagreeable and dangerous to work in.

In due course, the exact locations of the main ring and the incoming beam line were defined, and using optical and radar range finders a series of reference points was established, whose positions relative to each other and the existing CERN

95

Jean Gervaise (**centre**), *with members of his team, was responsible for surveying the site and establishing the coordinates from which the tunnelling machine could be guided. Before starting on the SPS site, in doing the mensuration for the ISR, he had made large-scale surveying into a science. The number of measurements and calculations required to fix with the necessary precision the components of the SPS were of the same order as those needed to prepare the maps for a small country.*

system were measured with a self-consistency of ± 5 mm. The depth chosen would leave at the shallowest point a roof of 11 m molasse topped by $6\frac{1}{2}$ m moraine, whilst at its deepest the tunnel would be some 65 m underground.

It was decided that the plane of the machine should be horizontal relative to a vertical line at the machine centre. Such a definition might seem unnecessarily pedantic, until the implications of the machine's size are taken into account. All verticals meet at the centre of the Earth, so at points on the Earth's surface separated by 2·2 km they are diverging at an angle of over one minute of arc—not a great deal if it is the carriageway of a road we are considering, but quite enough to wreck the carefully lined-up magnetic fields that guide the protons during their 150 000 turns in the accelerator. To correct this divergence of the verticals, each magnet would have to be tipped in by about half a minute relative to its local vertical, or in other words the outer edge of the magnets had to be cocked up by about $\frac{1}{5}$ mm with respect to their local horizontal. Nor is this all; around a perfect circle all points lying in the theoretical horizontal plane would be at the same level. But the long straight sections cut the corners, and appear to dip in the middle about $1\frac{1}{2}$ mm, which is ten times the error the machine can tolerate. There presents itself,

The design offices and development laboratories of the SPS were built in the form of three crosses beside a large hall, in which major components could be assembled and pre-installation testing carried out. The cross construction, apart from allowing adequate space for parking and giving plenty of light into the buildings, played an important role in establishing communication between the members of the different groups, who in the course of their work naturally meet in the central landings. The far building contains a canteen, post office and bank. The height of the buildings was limited to 10 m, corresponding to the height of the encircling trees.

therefore, this paradox of a device which because it must be co-planar cannot be all at the same level, and which is leaning in at all points on its circumference.

Before much headway could be made on the civil works, the construction boss (known more soberly as the Head of the Site Installations Group) had to take up his appointment. Not for the SPS a traditional seasoned muck-shifter, but a 'general engineer' from the Saturne accelerator at Saclay in France—a compact, hearty dynamo, bouncing with energy and good cheer, and packing a back-slap liable to put the unwary on the floor. Robert Lévy-Mandel was a member of the 300 GeV Machine Committee in the Magnet System working group. Now came the great change. His was the massive and complicated job of keeping under control the multifarious activities that would range from hewing out the tunnels and constructing the offices, laboratories, and assembly halls to choosing the colours of the shed doors.

The most urgent need was for a suitable work place for the design teams, reasonably near an assembly area, and where the various groups could be moulded into a closely co-ordinated entity. To begin with, they were housed in prefabricated huts at one end of the existing site; but it was clear that as recruitment progressed these would start to burst at the seams and the waste of time in moving to the scene of operations would be a source of irritation, especially as the interconnecting tunnel under the road had not yet been constructed.

In the new laboratory area, three buildings, three stories high, were built in the form of a cross. Seen from the air, the three crosses have earned the laboratory the title of 'the graveyard', but, apart from their economy of line and simplicity of construction, their shape has had a subtle effect on the life of the project. Although it is possible to enter or leave the building by any of the stairs at the ends of the wings, the route commonly used—either for coffee or to visit another section—was by way of the central cross-roads, where invariably others were encountered doing the same thing. It was impossible not to be aware of other people, other sections, other problems, and a sense of common purpose was generated which had much to do with this simple architectural feature.

Calls for tenders were sent out to European companies for the buildings, and for an assembly hall of 11 000 m^2 surface area, in which components could be assembled and tested before installation. A firm principle was established that all surface buildings should be functional and unobtrusive, of simple but pleasing architecture, and economical to construct. CERN indicated how this might be achieved, but left to the building industry a great deal of liberty in the choice of constructional techniques. In the adjudication, a French company was found to have presented the best offer from the point of price and technical specification.

The assembly hall was a plain, metal-clad construction. The pastel shade of the painting and the pleasing rhythm of the cladding and penetrations established the patterns for all the auxiliary buildings. When the construction mess was cleared up, grass sown, and trees planted, the whole laboratory complex melted smoothly back into the countryside. Its position, although largely conditioned by the siting of the synchrotron ring, had been carefully tucked in behind existing thickets, and when the Director-General suspected that not everyone on the site was as environmentally conscious as he was himself, he gave orders that no tree was to be cut down without his personal sanction. The original design for the ejected beam line to the north had to be changed because the number of trees that would have to be felled had not been kept to the minimum. With this firm insistence from the top, the laboratory planners and contractors soon assimilated the idea that care for the environment was more than just a slogan.

Actual site work began towards the end of 1971, long before all the land had been acquired, and even before the conclusion of the agreement with France

5/Down to Earth

Site work began before the end of 1971, well before all the formalities associated with the acquisition of the land had been completed, but by the end of the following year all the problems on the French side had been resolved, and the formal lease agreement was signed at a modest ceremony in the mairie of Gex, the principal town in the adjoining French territory. At the same time, the lease agreement concluded in 1965 for the extension of the laboratory for the ISR was amended to bring it into line with the new formula designed to cover a distributed group of international islands connected by public roads.

covering the legal status of the Organization in that country. That agreement was eventually ratified by the French parliament towards the end of 1972 and at about the same time, the lease agreement for the new land in France was signed. The *contrat de superficie* (for which there seems to be no tidy translation in English) with the Swiss Government was not signed until December 1974.

In spite of the great hurry to get started and the large number of owners (600) who were to be affected, no expropriation orders were needed, and all the land

Europe's Giant Accelerator

was acquired by friendly negotiation. In certain instances, these negotiations were a little protracted, particularly on the Swiss side where there is considerable scope for playing the Canton off against the Confederation, but the acrimony generated was minimal, thanks to the patience and skill of the competent local authorities and the tolerance of the land-owners concerned. Many have sold their land, but keep it just the same, and pay a far from exorbitant rent. In France, the rent collected is paid into a fund managed by a joint committee of CERN and local authority delegates, and used for the improvement of the environment. It is clear that the new research laboratory and the restrictions on future construction have guaranteed the continuing rural nature of the locality more assuredly than a vulnerable green belt plan might have done.

In the call for tenders for digging out the main ring tunnel, it was specified that it was to be bored rather than excavated by blasting and shovelling, and that contractors must already have had experience of using a boring machine in similar rock. CERN could not afford to be a guinea-pig, nor to find itself in the position of a project in Spain where the boring machine got stuck in the tunnel it was digging and resisted all efforts to get it out.

It was essential to avoid the use of explosives as far as possible to limit the damage to the surrounding rock structure, a consideration also appreciated by the local inhabitants. In the connecting tunnels, and in those parts where the tunnel cross-section was different from the standard circle such as in the enlarged straight section adjacent to the six permanent access shafts, mobile head diggers were employed to the maximum extent, but it was impossible to avoid some blasting, for example to clear rock falls, and for a few months the neighbourhood shuddered from time to time to the muffled thump of underground explosions, limited to the hours of daylight out of regard for the repose of the locals.

The successful tender was put in by a consortium of companies from France, Italy and Switzerland. It was proposed to employ a Robbins full-face boring machine, which would grind its way through rock, dragging behind it a long train of auxiliary plant that would dispose of the spoil and reinforce the tunnel walls as it went along. The tube cut out would be 4·8 m in diameter, which, after lining, would leave a clear aperture four metres in diameter. This would just leave space for the accelerator components, with a passage to the side along which replacement units could be moved in or out.

The underground navigation of this strange craft, which was to burrow its way along a complicated path for nearly seven kilometres with never a sight of the surface, was a fascinating problem, far from academic in its significance. Because the tolerances were so tight, any deviation from the ideal line of more than the odd

5/*Down to Earth*

As work on building the central laboratory area continued, a start was made on the construction of the main ring tunnel. Although this was to be bored by a special machine, two shafts down to the desired depth had to be dug and a section of tunnel linking the two prepared in which the machine could be assembled. Here, work begins on the preparation of the first permanent access shaft situated just inside the line of the main ring.

few centimetres would be very costly to correct. The system of tunnel wall reinforcement depended on the section cut out being circular, and if it was necessary to gouge out a bit more on the side then this would hold up progress, would require non-standard reinforcement techniques, and all the carefully thought out system of automation would break down. Should the drift not be discovered until the boring operation was completed, the loss of money and time in making the work good would be great indeed.

In these circumstances reliance cannot be placed on most of the usual aids to navigation; in particular, the magnetic compass is not able to give the accuracy

Opposite

Eventually, six such shafts would be dug out round the ring, connected to the ring by a short horizontal gallery entering at an angle of 30°, to minimize the radiation that would stream in once the accelerator was in operation.

*Directly over the incoming beam line a service shaft (PGC) was cut that would be plugged once the tunnelling work was complete. Between the PGC and the gallery leading from the first shaft, a straight section of tunnel was dug out, using as far as possible mobile head diggers, which gouged away the rock with a revolving head, on a pivoted cantilever arm. By the gallery entrance (**right**), an enlarged cavern was prepared and a start made on the 4·8 m diameter hole (**centre**) into which the cutting head of the boring machine would be inserted.*

needed, because of the local variations in the Earth's magnetic field. One completely constant reference is the axis of the Earth's rotation, which appears the same everywhere. A gyroscope has the tendency to line up with this axis, and with the aid of a gyro-theodolite a given direction could be transferred down one of the vertical access shafts into the tunnel with great accuracy. Reference points had already been established on the surface at the access shafts, which also allowed the position of a point at the bottom to be defined with precision.

To start off, two shafts were dug down to the tunnel level, one vertically over the eventual incoming beam line, from the PS called the *Puit Génie civile*, or PGC, and the other PP 1, the permanent access shaft, near the long straight section No. 1, where the protons would enter the SPS ring. These were dug out largely by hand

The coordinates established on the surface were transferred down the shafts to the working area below. Here, the operation is being carried out at the PGC beside the lift shaft built to take two wagons in which the spoil was brought to the surface, where it was dumped before being spread on a nearby site, by now covered with top soil and grassed.

Opposite, top

Conditions in the tunnel for precision survey were less than ideal. Near the access shafts the instruments had to be protected from dripping water—a conventional umbrella serving for the purpose; although in the main ring conditions were for the most part dry, equipment almost entirely blocked the view, and the geometers had to contend with dust and railway lines laid on uneven sleepers underfoot.

Opposite, bottom

Parallel to the main ring work, a start was made on preparing the sloping tunnel that would carry the ejected beam line to the north area. This was dug through the moraine, which is treacherous and of highly irregular texture crossed by small water courses. Until the molasse was reached, the going was inevitably slow, as local falls required emergency shoring and reinforcement of the roof before further work could proceed.

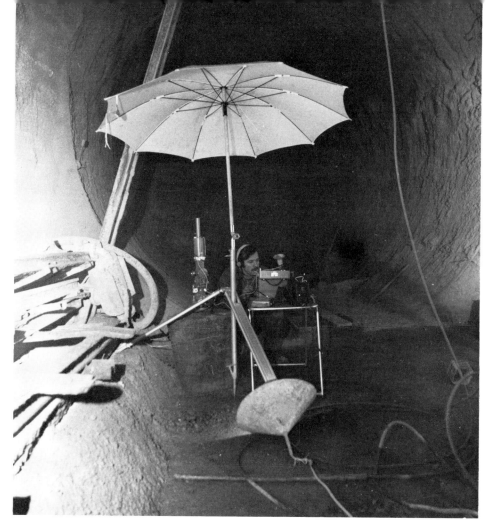

The head of the Robbins boring machine, armed with 32 cutting disks and weighing 60 tons, is stood on edge ready to be lowered the 46 m down PP 1 to the gallery below.

Once safely below, the delicate job of inching the head into the hole prepared to receive it was begun, and the components which make up the boring train attached. Round the periphery of the head were scoops which picked up the rock chippings and fed them back to a conveyor belt for transport to the waiting wagons.

5/Down to Earth

(and pneumatic drills), lined with sheet steel, and reinforced by steel hoops, after which low-pressure concrete was injected behind to fill in the pockets and provide a water seal. Once down to depth, an internal concrete lining was poured behind a sliding shuttering from the bottom upwards, and the lower section sealed to the molasse by injecting high pressure concrete at the level where the molasse and the moraine meet.

At the bottom, once a good space had been cleared, mobile head diggers went into action to dig out the tunnel connecting the PGC to the access gallery leading from PP1 to the main ring. The 9 m diameter PGC had been fitted with a 17-ton goods lift, capable of taking two wagons, which carried the spoil away from the diggers and dumped it on a pile for subsequent batch removal to a disposal ground. PP1 was only 5 m in diameter, and people went down and up in a bucket on the end of a cable. This shaft would just take the head of the Robbins machine.

Assembling the boring train, and setting it into the start position was itself no simple feat. The head alone weighed 60 tons, and once lowered down PP 1 had to be inched along into the cavern dug out at the end of straight section No. 1, then eased into a hole prepared for it at the beginning of its burrow. Behind it was assembled the drive unit, and behind that again the rest of the train, making up a complete machine over 60 m long. Driven by four 110 kilowatt diesel motors, the head with its thirty freely rotating steel cutting disks was carried on a shaft which forced it against the rock face with a thrust of 400 tons. Two massive steel chocks locked against the tunnel walls anchored the drive unit in place, and a pedestal down on to the tunnel floor took the dead weight. After the head had advanced about a metre, the chocks were released, and the drive unit moved forward ready to start another stint.

Direction was determined by regulating the angle the drive shaft made with the drive unit, but the machine driver has to know where to go! On the outer wall of the tunnel, every 32 m a stout bracket carrying a standard socket was fixed, and its position measured. Into each socket in turn was mounted a laser, set to point in a calculated direction. Its beam illuminated two grids fixed one in front of the other on the drive unit, and visible from the cab of the driver. He was provided with a chart which showed him where the laser spots should be shining at any moment, and it was up to him to follow the line indicated. Every 32 m the laser was moved and reset with reference to the original point and line to ensure that errors would not be cumulative. Out in the open, with plenty of room to move around, it all seemed straightforward, but these fine measurements had to be made in the narrow and dark confines of a tunnel of circular section, almost completely

*Behind the head was the drive mechanism, which was braced rigidly in position by hydraulic rams pressing on the tunnel walls. The head was driven forward for about one metre, after which the rams were withdrawn and the whole train advanced ready for the next thrust. The direction and cutting rate were controlled from a small cabin (**right**) underneath the conveyor belt carrying the spoil.*

blocked by the train over the last 100 m, with underfoot the perpetual nuisance of the train track and its wooden sleepers spaced at irregular intervals.

Chippings ground off by the boring head were swept into scoops, which channelled them into a conveyor belt system leading back through the train to the waiting wagons. These arrived loaded with reinforcement sections, and were lifted up to the top level of the train, where they were discharged and then lowered down in front of the wagons being loaded ready to take their turn.

On the surface alongside the PGC, a prefabricating plant had been built for the manufacture of vault sections, a ring of six making a complete circle, which would fit inside the raw tunnel and form the outer vault. The sections were cast in steel

5/Down to Earth

*On the surface close to the PGC and the dump for the spoil, a pre-fabrication plant for the manufacture of the interlocking concrete vault sections was built. These sections, cast in steel formers, were artificially aged in a steam-heated tunnel (**centre**) to give the concrete strength before installation.*

formers linked together as an endless belt on two levels, which passed through a heated tunnel where the concrete set and was artificially aged to give it immediate strength.

On the train, semi-automatic equipment picked them up and placed them as two sets of three against the tunnel wall, with wedges in between the top and bottom sets. These forced the sets together and jammed them against the tunnel walls, while concrete was run between them to fix them permanently in place. If the rock structure was of doubtful quality, the walls could not be left unsupported over the 20 m between the cutting head and vault placing unit, and the machine was equipped with a device for putting in steel reinforcing hoops just behind the head.

109

The reinforcing sections were transferred to the tunnelling machine by the wagons and placed by the machine against the wall, six forming a complete ring against the wall. The rings could then be grouted permanently in place.

These could not always be removed afterwards, and it was necessary to leave them in place and complete the strengthening with steel mesh and sprayed concrete. One more function performed by the train was to lay its own tracks as it went along.

In the beginning, obviously, progress was a little slow and hesitant, but as the driving team gathered confidence the rhythm picked up, and before long the programmed 20 m per day was being achieved. Out of the straight and into the curve the steady claw to the right began, and soon the train disappeared for the first time round the bend and was lost to sight. It was then on its own for the next 1200 m, and fingers were crossed that it was aimed in the right direction.

Came the day when it was due to emerge into the cavern at long, straight section No. 2, that had been carved out starting from the second shaft and into

5/*Down to Earth*

Eighteen months after the head had been lowered down the first shaft, the great day came when it broke through into the section it had started from, having completed a nearly 7 km circuit with an average precision of about 2 cm; an astonishing piece of subterranean navigation. Over the heads of the crowds assembled to witness the event can be seen, on the right, the tunnel for the injection beam line which passes under the PGC.

which had been transferred a new set of reference marks. An anxious crowd gathered as the rumbling the other side of the end wall got louder—showing at least that the Mole, as it was known, was somewhere near where it ought to be. Finally, the rock face began to crumble, and to cheers the revolving snout broke through. Not long afterwards, when some of the dust had cleared away, Gervaise proudly announced that it was within about 2·3 cm of the ideal line. It looked then that it would be possible after all to close the circle, and the men on the Mole would not be doomed to cutting an endless spiral.

For most of the run, the rock stayed good, so that no-one was prepared for the moment when high pressure methane began to hiss out of a fissure into which the head had cut. The tunnel was immediately evacuated because of the danger of explosion, and it was several days before the emergency ventilation measures made the tunnel safe to enter and the machines could re-start. A little oil was also encountered, but not enough to cause serious problems, nor to make CERN contemplate turning itself into an oil producer. The circle was closed at the end of July 1974, 16½ months after the Mole had set out on this circumnavigation under the Franco-Swiss border. Subsequently, a detailed measurement of the position of alternate vault rings showed that over 99 per cent were within 10 cm of the ideal line—a tribute to the efficiency of the Robbins machine and the skill of its drivers.

By the time the Mole had completed half the ring, all the spoil had to be run back through the 3.5 km already dug, and no other work of any importance could go on inside. Moreover, there were many more operations to perform before the tunnel could be considered ready to receive equipment. Consequently, the factory for the fabrication of reinforcement sections was moved to the head of the shaft No. 4 (PP 4), which was equipped with a lift similar to that installed at the PGC. For the second half, communication with the Mole was via PP 4, and the finishing work on the first half could start.

Contrary to what might have been expected, it began at PP 4 and worked back towards PP 1. The reason for this was that PP 4 would be fully occupied in tending the Mole, so all supplies would have to be passed up to the work areas from the PGC. In addition, once the floor was poured there could be no access across it, so this had to be done last at the point farthest away from the supply line. A consequence was that all the machines engaged in the finishing work had to be designed to straddle the supply line, leaving a free passage in the middle for trucks to go through.

The sequence of operations was highly automated and designed to proceed at a constant rhythm, each team retiring round the ring at the same pace. When all teams were in action, they spread out over an arc of some 400 m. Bathed in little islands of light, throwing grotesque shadows into the gloom around them, they worked away to the constant clatter of their machinery, punctuated at irregular intervals by the percussion of exploding nail guns, and the rattle of passing trucks.

The first operation was to spray the joints between the reinforcing ring sections and fill the spaces behind with weak concrete to spread the load on the rock. Channels were left periodically to guide any seepage of water into drains at the bottom, from which it could be pumped away. The hydrostatic pressure that can

Many more operations were needed before the tunnel became habitable: the drains had to be prepared, the steel lining nailed to the walls, the inner vault poured, the inner and outer vaults grouted together under pressure, the equipment cavities in the floor prepared, the floor poured, and then the whole area cleaned and painted. But section by section, the tunnel was handed over to the measuring and installation teams—a clean, dry building, 4·0 m across, curving away into the distance.

build up under a column of water 50 m high is over 5 atmospheres, thus not to be trifled with.

Even so, water-proofing was still necessary, and this could not be done with heavy plastic sheets welded together in those tunnels where the radiation levels might be high. Radiation from lost beams or equipment made radioactive by stray protons causes plastics to become brittle and fragile. Although the levels of activity expected in the main ring are quite low, and there will be no problem of personnel entry when the accelerator is shut down, it was felt to be too big a risk to use plastic water-proofing there. Instead overlapping steel sheets were nailed to the walls before an inner vault was poured.

The shuttering machine which defined the form of the inner vault was an ingenious device. Made up in three sections, the end one could be collapsed just enough to let it pass through the other two to take up a position in front. While the concrete was setting and growing in strength supported by the others, a new section could be poured and left to set in its turn. As the finished tunnel diameter was only 4 m, and, as already made clear, the machine had to straddle the supply line, the amount of room to play with was minimal, and there were quite a few anxious moments when it seemed that the tolerances were a bit too tight.

When the new vault was strong enough to stand the strain, concrete under high pressure was injected behind it, the pressure being maintained until the concrete had taken. In this way the gaps between the steel sheets were sealed, and the outer vault was strained out with a uniform force against the surrounding rock while the inner vault was compressed. Subsequent movement of the rock might alter the loading, but the inner vault should always remain under compression, so not be subject to cracking.

When the vault had been poured, the drains put in, and everything cleared, the rails laid down many months before were taken up and fixed to the walls at a constant height. Along them ran the floor pouring machine, which left behind a clean, open stretch suddenly become quite roomy, quiet and permanent. There were still the rails to dismantle, the instrument holes to tidy up, the walls to be painted, the floor to be sealed and vacuum-cleaned, but the essentials were complete, and installation work would begin as soon as one sextant of the circuit could be sealed off from the retreating army.

6/Guiding and Focusing

"Guide the beam of protons round the 7 km ring at almost the speed of light for 200 000 turns with a precision of a few millimetres". That was the essential specification given to Roy Billinge, a dynamic extrovert—and gastronome—originally from the Rutherford Laboratory, who had gone to CERN in 1966 to work on the design study of the 300 GeV programme. Characteristically impatient with the waiting, he had leapt at the opportunity offered by R. R. Wilson to take part in the construction of the Batavia machine in the USA. There, he led the group which built the 8 GeV injector synchrotron and the first beam lines, and completed the job by the time the decision on the SPS was made.

Uninhibited about pushing through his own ideas, or taking up arms on any technical subject that arose, he returned to CERN with a detailed knowledge of the American project and an admiration for its leader. He was convinced, also, that a change was needed from the traditional practices of magnet building in Europe. Many doubts were expressed when, for example, he eliminated mica insulation from his magnet designs, but he was emphatic on the need for keeping the cross-section of the magnets to a minimum, confident that the field specifications could be met.

The protons have to travel a million kilometres inside the ring, three times the distance between the Earth and the Moon. Moreover, as all the protons carry the same positive electric charge, and so mutually repel each other, they must continually be persuaded to stay together to form a thin beam occupying the least possible space in the vacuum tube in which they circulate. Every millimetre extra

Right
Roy Billinge returned to CERN from the USA to take charge of the magnet system of the SPS, the biggest single component of the machine which guides and focuses the beam of protons as they are accelerated round the 6·9 km ring for 150 000 turns, a distance equal to three times the distance between the Earth and the Moon, over which a precision of the order of one millimetre was required. He held strong views on magnet design, which did not conform with CERN traditions.

in the dimensions of the tube means more iron and copper in the magnets which guide them, more expense, and also more power in operation. As a further complication, the design had to be capable of being extended at a later date to take account of the decisions Council would make ultimately. In the first instance, only half the available circumference of the ring was to be filled with bending magnets, but, later, magnets of a similar or different type would be added to obtain the higher energy agreed upon. As for the focusing magnets, these were to be designed from the outset with a capability of 400 GeV.

The focusing magnets are spaced regularly around the ring, focusing alternately in the vertical and horizontal planes. There is no practical way of making a magnet which will squeeze the beam strongly all round. A quadrupole magnet, with four poles, alternately N and S, set at $45°$ to the horizontal plane, can exert a compressive effect in either the horizontal or vertical plane, but not in both at the same time. As the protons move outwards in one plane, the force gets stronger, with a focusing action in one plane and a defocusing action in the direction at right angles. There is no way of getting round this; with more poles, the same thing happens but in more planes. Happily, if magnets focusing in perpendicular planes are alternated, the net effect is that the beam is confined within a narrow envelope of regularly changing section.

Protons going right down the middle of the focusing magnets feel no effect as the field there is zero, but few do this. The majority, as they speed round the ring, are swinging, first to one side, then to the other, and up and down, the profile of the bunches changing with a regular rhythm. If we could take a ciné film from the side and from above of a particular bunch, we should see it contracting as it left one quadrupole, and expanding as it left the next, contracting and expanding alternately, and the two films would be out of phase with each other. However, should any individual proton find itself repeating, in the same places on successive turns, the same oscillations, either side to side or up and down, a resonance would develop; very quickly the amplitude of the swing would build up to uncontrollable proportions, and the beam would be lost. The number of oscillations a proton makes in one turn is a very critical quantity, and is called the Q-value, and it depends on the strength of the focusing field in relation to the energy of the beam. In the SPS, the value chosen is around $27\cdot5$, with variations of about $0\cdot1$ on either side acceptable. This avoids the resonance values at which the beam's oscillations would build up. If we could film the orbit of one proton, we should find it behaving differently on every turn; otherwise it would soon disappear from view.

In addition, not all the protons have the same energy. The spread is small, but it is enough to change the orbits on which they travel. The more energetic ones

6/Guiding and Focusing

In both the PS and the ISR, guiding and focusing is done simultaneously in C-shaped magnets with projecting pole pieces of complex shape (to combine the guiding and focusing function) around which the energizing coils are placed. With these magnets, both the PS and the ISR have given long and reliable service.

swing out wider than their fellows; to them the focusing fields seem a little weaker, so there is a spread of Q-values over the protons in any bunch. This is a characteristic which makes for even tighter tolerances in the focusing system during acceleration, but which can be exploited when the time comes for ejection.

Europe's Giant Accelerator

*In the SPS, the function of bending and focusing is separated. The bending magnets are required to produce a field in the vacuum chamber uniform and reproducible to one part in ten thousand. The steel core which surrounds the chamber is split into two halves, and the chamber is compressed between the two halves. The energizing coils are of hollow copper bar, through which cooling water circulates, insulated by a wrapping of resin-impregnated fibre-glass (**not shown**). Altogether in the main ring are 744 bending magnets, containing 12 000 tons of steel and over 1200 tons of copper.*

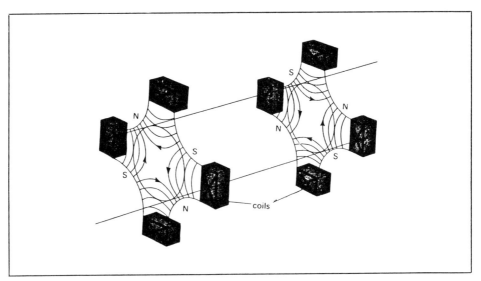

A quadrupole magnet, with alternate north and south poles separating the coils, shown in the drawing, produces the shape of field indicated. The force exerted on the circulating protons is at right angles to the field lines and is zero in the middle. In the first magnet the field squeezes the beam horizontally, but expands it vertically, while the reverse happens in the second magnet. By alternating the orientation of the poles of a succession of magnets, a net focusing action results.

6/Guiding and Focusing

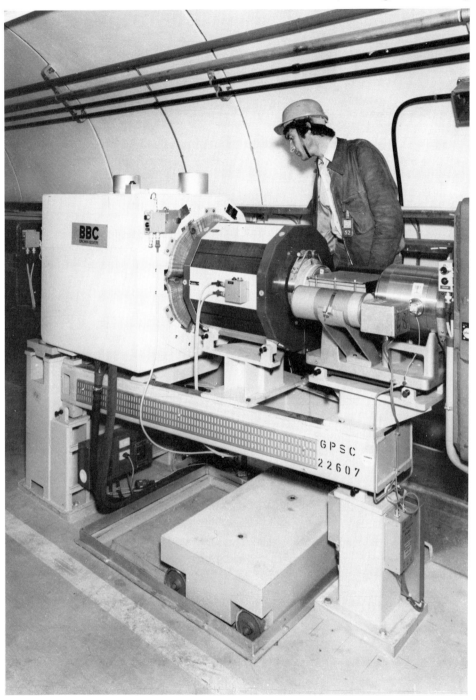

In the main ring system, in addition to the principal bending and focusing magnets, are correction magnets. Multi-pole magnets (**left and centre**) are followed by a simple correction dipole, followed again by a beam position monitor which records the position of the bunches of protons with respect to the walls of the vacuum chamber.

The Q-values of the beam are measured by deliberately kicking the beam with a magnetic field to a few millimetres away from the equilibrium position, and measuring the frequency of the resultant oscillations. The value found can then be adjusted if necessary by varying the field in the focusing quadrupole magnets.

For the bending magnets, the ideal is a perfectly uniform field, which is the same for all the magnets along their complete length, across the entire area of the vacuum tube. The field must remain uniform to three parts in 10 000 throughout the ring, whatever the current flowing through the energizing coils, so that as the field rises from the beginning of the acceleration cycle to the end, it is everywhere identical in the vacuum tube inside the magnets.

A particular problem arises at the lower field values when the current is small, because the residual effects in the magnet steel then assume an especial importance. Steel capable of producing a high magnetic field does not demagnetize completely when the energizing current is switched off, and the exact value of the remanent field depends critically upon the steel composition. CERN already had considerable experience in coping with this particular problem, and its pioneering work in the development of steels with a very low carbon content has had an influence far beyond high energy physics. Steels developed essentially for use in high energy accelerators are now to be found in compact small horse-power motors, and, surprisingly, in enamel ware. Washing machines and bathtubs on the market are better and cheaper because of the demands of accelerator builders. There was no need to look further: the steel adopted for the SPS magnets was the same as that specified for the ISR, whose performance was well proved.

The simplest design of bending magnet with the easiest access for the vacuum tube is the C-shaped magnet, where the coils are mounted round the projecting poles well away from the gap between them. In order to obtain good uniformity at high fields, the magnets must be very large: the penalty, in weight of iron and copper, and in operating costs, could not be tolerated in such a large installation as the SPS. But costs are not the only consideration. While the magnets are being energized throughout the whole cycle, the copper coils heat up and must be cooled by circulating water through them. There are limitations on the quantity of water that can be pumped through, and on the temperature rise that is acceptable. Even more significant are the restrictions on the peak power that could be taken from the mains supply, and on the stored energy in the magnet system. Much of the energy which goes into the magnets when they are charged is stored as magnetic energy, which comes flooding back into the source when the power is turned off after the protons are ejected. If the whole of the region around CERN was not to find its electricity supply pulsing up and down to the rhythm of the SPS, both peak power and stored energy had to be kept to a minimum. These parameters also would determine the rate at which the accelerator could be cycled, which in turn determines the output of the machine in terms of number of protons accelerated.

An H-frame magnet (confusingly, the C describes the shape of the iron yoke, the H the aperture left in a rectangular iron block), which is in effect a double C, is better than the C, but it still contains a lot of unused field, and in consequence is wasteful. A way round this is to use a window-frame construction, where coils fill the two sides of a rectangular aperture in a rectangular block. However, to achieve the necessary uniformity, the pure window-frame magnet must either be made very wide, which again is wasteful, or with few turns, which once more raises power consumption problems.

The Magnet Working Group had preferred the safer H-frame design, but Billinge insisted on a mixed system—an H-frame magnet with additional turns on the axis, accepting the problem that the alignment of the inner coils would be crucial, and the vacuum tube would have to be integral with the magnet. As the profile of the beam changes continuously, the aperture required also changes, but on economic grounds the number of types of bending magnet was limited to two. Those on either side of the (horizontal) focusing magnets, where the beam is thinnest, have a flat aperture, whilst those on either side of the de-focusing magnets (focusing in the vertical plane) have a squarer section. The average beam width is always bigger than the height, because of the spread in orbit radius. The length was fixed at a little over 6 m, the cores (or yokes) to be made in two halves, of steel laminations $1\frac{1}{2}$ mm thick, stacked face to face. The coils were made of hollow copper bar of the order of 3 cm square, with a central hole 1 cm diameter, insulated by fibre-glass tape and potted in resin. Wired in series, the complete ring has a resistance of 2·86 ohm, and can carry a peak current of about 5000 ampere.

For the focusing magnets, where symmetry is of prime importance, and where different shaped vacuum tubes can be inserted into the cross-shaped gap between the poles, it was decided that a single design would be optimum, 216 being required to go round the ring. (The regular pattern is repeated even in the straight sections, as the protons can never be allowed to deviate more than a certain amount from the centre line). Each quadrupole is 3 m long, with windings of hollow copper bar of 1·7 × 2·5 cm section, with a 0·65 cm diameter hole down the middle. Peak current is 2000 ampere.

Designing is one thing, manufacturing is another, and this was a programme that demanded great reproducibility over a long series run. The very length of the bending magnets, the need to assemble them round the vacuum tube, and to make adjustments to the position of the centre conductor before welding them up, checking, testing and installing, pointed to their manufacture at CERN. The quadrupoles, shorter, with coils set back from the pole faces, could be made outside CERN to dimensional specifications, with the calibration done at CERN. Similarly, the various correction magnets, which would allow the beam to be steered into the exact line desired, and the multi-pole magnets, which compensate for eddy currents and interference between the vertical and horizontal focusing fields, could be given mechanical specifications, and were small enough not to create transport problems. These, too, could be ordered complete from industry.

Test models were made at CERN to check the design, and when these were satisfactory tenders were sent out to European industry for the various components. Contracts went finally to a British company for the half-cores of the bending magnets made from Belgian steel, to another British company and a French company for the coils, yet another British company for the quadrupoles, and the various correction magnets were ordered from companies in Switzerland and France. Arrangements were made for regular inspections to be carried out in the factories, and it was not unusual for CERN staff to find themselves obliged to sort out not simply the technical problems which arose, but also managerial problems of keeping the lines running and organizing the necessary quality control.

Before assembly work began, the decision had been taken to go straight to 400 GeV by filling the ring with bending magnets of the type already designed; so the job was to manufacture 744 for the main ring, plus a few spares and those needed for the beam lines (the lines for leading protons into and out of the ring), making nearly 900 in all. Options included in the original tenders submitted by the suppliers were taken up by CERN. However, the British coil manufacturer felt unequal to the task, and all the additional coils were made by the French counterpart. No other manufacturer could step in at this stage and match the price, or go through the expensive process of jig-making and learning the procedure. The frustration felt by the Government representatives who had worked hard to encourage industrial participation by the UK in this large enterprise, after a very poor showing at CERN in previous projects, was intense.

At the rolling mill, samples of the magnet steel strip were taken at the beginning, middle, and end of each coil, and the physical properties and coercivity (which characterizes the remanent field) measured. This allowed the stampings to be shuffled at the core maker, so that, on aggregate, each magnet would have the same remanent field characteristics, a necessary procedure as every coil had slightly different properties, even though the spread permitted was very small. To cater for any asymetry in the die which made them, the laminations were reversed at regular intervals along the length. A complete half core was assembled by

6/Guiding and Focusing

Components for the bending magnets were made by European industry under contracts awarded according to the rules of CERN to the lowest bidder able to meet the specifications and delivery dates. Half-cores, six metres long, made up from lamination $1\frac{1}{2}$ mm thick, whose orientation changed regularly along the length, were set in jigs and made ready to receive the coils and vacuum chamber.

The coils with bend up ends, made to the limits of accuracy possible on such lengths, were then swung into position and measured to find the exact location of the copper in the inner coils, as the placing in the core had to be controlled by the copper rather than the rougher exterior of the insulation.

stacking the number needed to arrive at the correct weight, so taking care of small changes in thickness; then compressing them inside a frame in a massive press until they made up the right length, following which the end plates were welded on. In this way, every half-core was as identical to its fellows as it was possible to make it.

At the coil factories, the messy, difficult job of making the insulated copper loops got under way. All manufacturers of this type of product run into difficulties, but both CERN contractors had supplied coils for the American machine, and it was just a question of grinding away until the techniques for meeting the new specifications had been properly mastered. Hesitantly, and with some delays, the first coils began to arrive at CERN ready for assembly. Meanwhile, production of the stainless steel vacuum tubes, in Italy, had been going on, and these, too, were arriving at CERN, where they were given a special cleaning before being stored ready for the assembly line. The vacuum specification is relatively modest for an Organization that has built the largest and most intricate ultra-high vacuum system in the world (the ISR), but there has, nevertheless, to be a strict control of surface finish of the tube and also of wall thickness, as the eddy currents formed in the wall depend on this dimension

The vacuum chambers of stainless steel were degreased and washed to ensure absolute cleanliness inside. The section and wall thickness were designed to withstand the imploding effect of the vacuum, while presenting the minimum thickness of metal between the poles of the magnet and the circulating beam inside.

As soon as the big hall on the new site was finished, two assembly lines were built for the bending magnets as well as test bays in which all the magnets could be given a complete check-over and their characteristics measured. Apart from the mechanics of the operation, the challenge was to mould a team of talented and imaginative people, of whom some were required to carry out a mass production job at CERN, while others toured the factories that were making the components. It was, on the whole, a happy group that was formed—certain members even a little too light-hearted for the more staid members of the Project, who found it difficult to believe that the meticulousness needed would not be compromised.

In the assembly process, the inner coils were first checked to find the exact position of the copper under the insulation, and the coil assemblies placed in the depressions in the lower half core. Spacers set them in the right position, following which the vacuum tube could be put in place, and the upper core laid on top. The two halves were drawn together by passing a current down the middle to set up a magnetic field with a compressive power of hundreds of tons. Still locked in this way, the assembly then moved into a jig which bent the middle upwards, just enough, so that when the magnet came to sit on its four corner feet it would sag

Europe's Giant Accelerator

Sputter ion pumps, here being tested in pairs before installation, are the main means of producing a high vacuum in the tube. Connected by quick release joints, they bear little resemblance to the more familiar mechanical pumps with their characteristic clatter, and are much more convenient than the old diffusion pumps, used on the PS. Inside the sputter ion pumps, a jet of accelerated ions entrains and ionizes the residual air in the system, which is then removed by electric fields.

into a perfectly straight line. A welder on each side (to even out the strains) made the marriage permanent by welding steel straps across the join. The feet could then be attached using distance pieces, specially ground to ensure that the magnet would sit perfectly level. Reference sockets were precisely positioned on top, and a coat of red paint finished the job.

Transferred to the measurement bay, each magnet was required to pass through a long series of checks and calibration tests. End shims were added to make the bending strength at mid-field exactly the same as the reference magnet, and the magnitude of the eddy currents assessed and compensated by adjusting the shunt resistance across the winding. The steel had been selected and shuffled on the basis of the remanent field, but there was still the problem of small differences in the

With coils and vacuum tube in place, the assembly was ready for the upper half-core to be lowered into position, and the two halves drawn together by passing a high current through conductors slid into the vacuum tube. At the same time, the assembly was forcibly bowed upwards in the middle, so that when the tie straps were welded on and the strain relieved, the assembly would sag into a perfectly straight line.

way the steel saturated as high magnetic fields were approached. The solution was to arrange the magnets in groups of four, so that when installed in the tunnel each group would have the same aggregate saturation properties.

The quadrupoles were made from the same steel as the bending magnets, stamped into quarter sections so that each pole could be assembled independently. They were shipped complete to CERN, where the field gradient was plotted, and the exact centre line relative to the pole faces determined. Using a laser beam to transfer the co-ordinates, the reference sockets were positioned to an accuracy of better than 0·1 mm.

Setting the quadrupoles in place in the tunnel is an operation even more critical than placing the bending magnets, as right across the section the field changes from

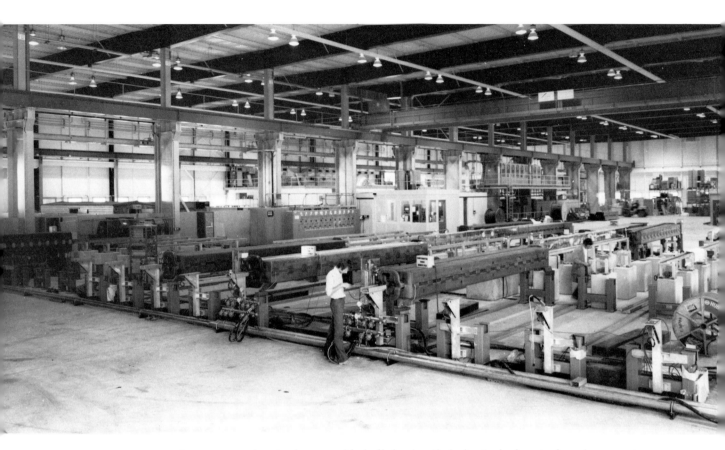

In a separate bay in the assembly hall the detailed physical, electrical, and magnetic characteristics of the magnets were measured and recorded. When they were installed in groups of four in the ring, the aggregate bending power of every group at all current values was as uniform as it was possible to make them.

zero along the centre line to a maximum near the poles. Displacement of a single quadrupole by a given distance results in a beam distortion five times as big, and a random error of the same size all round the ring gives rise to a distortion 100 times as big.

As soon as the first tunnel sextant became available, Gervaise and his team got down to the job of establishing the series of reference points every 32 m, close to where each quadrupole would be situated. These were levelled, and the eccentric mounting locked to roughly the right place. Using instruments developed at CERN, the distances between them, across and diagonally, were measured using a calibrated invar wire stretched to a standard tension.

This allowed the quadrilaterals formed by adjacent pairs to be determined, and these were put together by measuring the distance of each socket from nylon wires

The quadrupoles, each 3·3 metres long, were manufactured by industry and delivered complete to CERN. Here, they had to pass a stringent series of tests, and the exact centre line of the system established with reference to the locating sockets mounted on top, which subsequently would be used to position the magnets in the tunnel with an accuracy of 0·15 mm. The vacuum tube could then be slid in.

Manoeuvering the magnets into place in the tunnel was a delicate operation, for which special handling trolleys had been designed. Once, however, the techniques had been mastered, it became a routine matter, and a steady rate of five per day was maintained. There was little space to spare, however, and although personnel could just scrape past, the transfer of equipment was excluded.

Space was needed, also, for the measuring teams who were responsible for adjusting the exact location of the supports. When this had been determined with the magnet carried on hydraulic jacks, screwed distance pieces were set to take the load and locked immovably into place.

stretched between the sockets on each side on the outer walls, and diagonals set up from the sockets on the inner walls to those on the outer. The offset measurements were later checked by a specially designed laser technique. Feeding all these results into a computer allowed the position of each socket in a sextant to be determined with a self-consistency of 0·15 mm.

Subsequently, the sextants could be joined into the complete ring. Only distance measurements can be used for survey work over long distances. Optical systems (unless channelled through vacuum tubes which is clumsy and expensive) are excluded, because small changes in temperature along the tunnel would bend the light beam an indeterminate amount. The familiar mirage effect experienced when

6/*Guiding and Focusing*

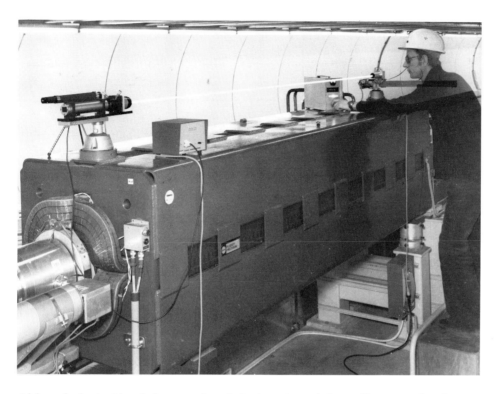

Although the inside of the tunnel and the location of the wall mounted, reference sockets had been determined by distance measurement, a check system based on a laser technique was developed. Chances could not be taken with a machine that was so big that all magnets had to tip inwards relative to the local vertical in order that all should line up with the vertical at the ring centre. Moreover, in the straight sections, they had to dip downwards relative to the local horizontal in order to remain in the same plane.

driving along on a hot day, which results in an image of the sky being seen in the distance in place of the road, would cause errors of an unacceptable magnitude. Admittedly, in general, the temperature gradients in the tunnel are much smaller than in the open, but the effect is there nonetheless.

Up in the assembly hall, the magnet stocks were growing, and space was becoming tight, but by November 1974 all was ready for the first magnet to be ceremoniously wheeled into the tunnel and lifted into place. The reference sockets on the magnets were lined up in relation to the wall sockets, the appropriate tilt of

Walking between sectors may be healthy, but it is also time-consuming. Because of the big distances involved, the staff was provided with electric trucks and cycles to move around the ring. The rule that people left from the same shaft that they had entered avoided the situation of finding all the transport parked at the bottom of one of the shafts.

0·2 mm across the width to arrive at a co-planar ring added, and the jacks and positioning screws locked up. In a short time, the installation had become routine, and magnets were being wheeled in at the rate of five a day. The forward programme seemed pretty comfortable. However, in what Adams later described as 'black January', misfortune struck. A technician reported that when carrying out a regular test in the tunnel he found that two magnets were showing insulation faults to ground.

One of the elementary tests made on the bending magnets after assembly was to check the insulation of the coils, but as these had been subjected to intensive tests at the manufacturers, including an underwater test, and again on arrival at CERN, it seemed unlikely that there would be any trouble on that score. The shock to everyone was profound, and there arose immediately the spectre of the American experience, where every magnet in the main ring at Batavia had had to be replaced, some more than once. Conditions at CERN were different. Learning from the American troubles, the insulation thickness in the SPS magnets had been increased, there was a greater clearance between coil and core, more care had been taken during assembly, and the test programme made much more rigorous. Installation had been done under dry conditions, whereas the accelerator building at Batavia had been constructed in winter from pre-cast concrete rings, and had been pouring with water when the frozen ground melted.

However, those who had opposed the innovations in the SPS magnet designs were not slow to point out that this was what they had predicted all along, and CERN should have stuck to the tried and tested methods of the past. It was useless to try to explain again that these would have brought their own problems and an unacceptable cost penalty; the sceptics would not be convinced. Pressure on Billinge was intense as other magnets, both in the tunnel and in store, were found to be faulty.

The Project was in near panic. The offending units were stripped down, and a minute examination started. The crucial clue came when Billinge was tempted to rub a discoloration with his finger and then lick it—giving rise to what later became the watch-word of this gastronomic group, "when in doubt, taste it". There was no doubt; acid was present, which Battelle Institute identified as phosphoric acid. A systematic search began to find out how it came to be there, while at the same time a study was made of the effects of acids on resin-bonded fibre-glass. Confirmation came that strong acids did affect the insulation properties of the fibre-glass composite, after a delay whose length depended on the concentration.

Acid contamination was then a plausible explanation. Finally the source was tracked down to a cleaning fluid introduced by the French manufacturer at a certain juncture to prepare the copper end connexions. Although these were thoroughly washed afterwards, some fluid stayed trapped under the coil supports, and remained in contact with the surface of the resin. Over the ensuing weeks, it crept insidiously down the glass fibres to create a conducting path from the coil to the core.

The problem now was to discover how many magnets were likely to have been affected, and how many would have to be re-made. All new coils made after that date were given an additional wrapping of insulating foil during assembly, and it was decided that 280 magnets, of which 100 had already been installed in the tunnel, would have to be broken down, checked, repaired where possible, and

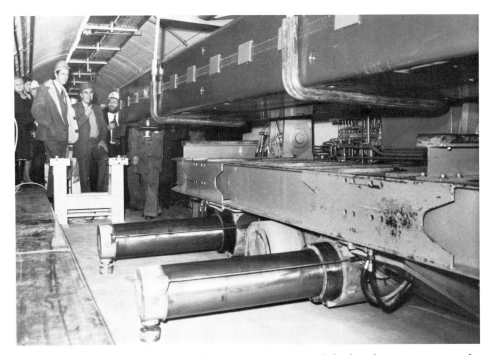

When insulation problems were discovered in some of the bending magnets at the beginning of 1975, it looked as if the schedule for the whole Project might be compromised. Rapid and incisive detective work revealed the cause, and the Magnet Group set about remaking the necessary number of units, while the installation teams got busy on a new schedule which involved removing as well as installing. To the undisguised satisfaction of Billinge, and the delight of Adams, the last magnet was moved into place on schedule in December of the same year. Five months would go by before it could be appreciated how good the system was.

then re-built. The load on the group and the manufacturer was considerable; the assembly line was put on over-time, the installation work was re-programmed, and the race was on to recover the time lost. Of course, the race was won; all the delay was wiped out, and in December 1975 the last magnet was taken down in triumph to its resting place, where Adams personally performed the last brazing operation connecting it to the power supplies.

The main ring tunnel was dug out by a full face boring machine (seen in the enlarged tunnel section prepared at straight section 6 before it began the last sector of the complete ring). Behind the head, which is forced against the rock face by a revolving ram, is a train of equipment for disposing of the spoil into waiting wagons, which arrive loaded with pre-cast concrete vault sections. The machine has devices for locking the vault sections in position, and for laying the rails on which it runs.

Direction of thrust on the head is controlled from a cabin. A laser on the tunnel wall directed along the train illuminates two screens visible from the cabin. The driver has a chart indicating where on the screens the spot of light should fall.

Each time the head broke through into an enlarged straight section, at the end of its traverse of more than 1000 m, the geometers were waiting to compare its position with the new coordinates transferred down the access shaft into the tunnel. Results were satisfying: for instance the distance off centre never exceeded 2.7 cm.

Final operation in the fabrication of the bending magnets: welders working in pairs, to ensure no distortion on welding the straps permanently holding together the two half-cores which surround the coils and vacuum tube.

After fabrication and an extensive programme of testing, the magnets were stacked in the assembly hall in classified groups, so that when installed in the tunnel each group of four would have the same aggregate magnetic effect.

Europe's Giant Accelerator

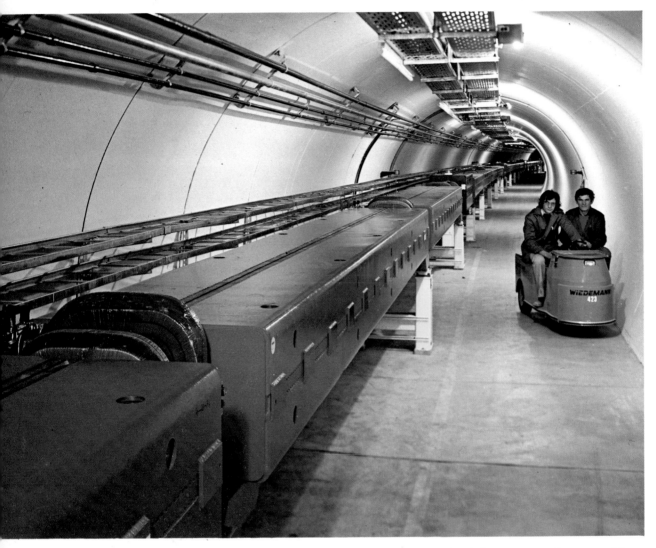

Magnets mounted inside the tunnel on the outside of the curve. Each group of four (red) bending magnets is separated by a quadrupole magnet (blue), a group of correction magnets, and a beam position monitor. On leaving a straight section, the scene repeats itself over and over again for 1200 m, a distance which seems much longer underground where all sense of position is lost.

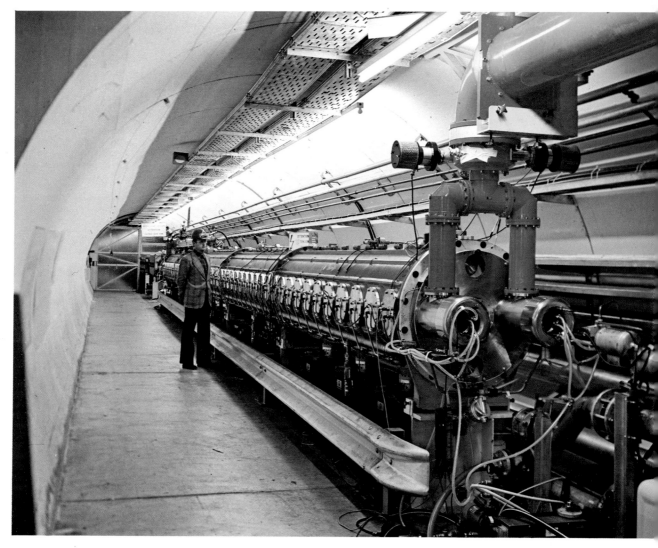

On each side of the entry from access shaft 3 is an accelerating cavity, able to give an additional energy of up to 1·2 MeV to the bunches of protons circulating on each pass. Each cavity is composed of five sections, in which a radiofrequency electric field is established. Power is pumped in from the downstream end (**right**), and the residual power emerging is absorbed in a water load (**upper left**).

Water to cool the closed circuit water system, which cools the machine components, is taken from the bottom of Lac Léman at Vengeron. At the junction of the motorway to Lausanne and the route du Lac, a pumping station and purification plant were built. These can handle a flow of 1·5 cubic metres per second, of which one third supplements the water supply of the satellite town of Meyrin near the CERN site.

Power is brought in on a 380 kV grid line from a French power station, 35 km away at Génissiat, coupled to the European grid network. Terminating the line are two 90 MVA transformers, one for the main magnet power supplies, the other for the rest of the site. Immediately beyond, and to the right of the sub-station, is the heating plant. Further to the right, the auxiliary building 3 (housing the power supplies for the accelerating cavities in the tunnel) and the control room. Above are the laboratory buildings.

Main control room with three identical consoles, equipped with coloured and black-and-white television screens, to display tabular matter, graphs, and minic diagrams; colours indicate the status of the information presented, or provide a contrast for setting cursor lines. The operator has a typewriter, through which he can give instructions in basic English; a ball for setting cursor lines, or indicating the parameter in which he is interested; and a knob for regulating values. Also, he has a touch panel through which he can call up machine components and computer programs. One operator can look after a machine with about 6000 adjustable values and about 10 000 situations to survey!

From straight section 6 appear vacuum tubes carrying beams of protons which generate particles destined for the west area. After leaving the accelerator, the protons can be directed on to a number of targets which serve the various experiments. The two on the left are for neutrino beams and the two on the right for hadron beams.

Europe's Giant Accelerator

Hadrons produced in one of two targets are directed into a 300 m evacuated tunnel, where they decay into muons and neutrinos. The tunnel is closed by a 2 mm thick titanium window, which must be protected when people are working in the area by a thick steel plate.

7/Power and Water

Peak power consumption of the main ring magnets of the SPS is 135 000 kW, of the same order as the average consumption of Geneva. In a town, however, because of the large number of users, the effects of switching on and off are largely smoothed out, and although there are mass movements as factories start up in the morning, or a popular television programme finishes and people switch on their kettles and hot plates, the rate of rise and fall in demand is relatively slow. Also, these mass movements occur at most a few times a day, and they can be anticipated by the generating stations. The SPS switches on and off once every six seconds and, not content with simply varying its power demand between zero and some maximum, it asks for a fair proportion of the power it has absorbed to be bought back again each time after the beam has been ejected and the magnets are de-energized.

When any sudden alteration is made to the voltage applied across a magnet, the magnetic field opposes the current change, and with a magnet the size of an SPS bending magnet, several seconds would elapse before the current settled down to its new value because of the very large inductance of the windings. In order to produce a current flow which rises at a pre-determined rapid rate, a much higher voltage has to be applied than is needed to maintain the corresponding steady current. And once the peak current has been attained, the applied voltage must be reduced quickly to the level corresponding to the steady-state condition. Again, while the current is diminishing, a voltage is generated tending to oppose the diminution, whose magnitude depends on the rate of change of the current.

After the protons have been ejected from the machine, opening the breakers in the coil circuits of the ring magnets would certainly cut the current flowing through them to zero, but with disastrous results, as the voltage surges induced would be huge, electrical break-down would follow, and the whole system would be destroyed. The energy stored in the magnets must be disposed of in a controlled manner, and it is clearly economical to return it to the source from which it came. If a single generating station was required to handle such a situation, it would be like a public utility having to cope every six seconds with all its customers successively switching on and switching off together, and, in addition, being obliged to take delivery of a surge of power from another company into its generating plant just when its own network had cut out.

Nor is this all. Most domestic loads are dominantly resistive, that is the alternations of the current are in phase with the alternations of the voltage. In this particular case, as the voltage and the current are rising and falling together at a frequency of 50 cycles per second (50 Hz) the current is very nearly proportional to the voltage. The two are in phase, and the so-called reactive power load is small. The power converted in the appliance at each instant is what the generator 'expects' and is designed to produce. For an idealized magnet with no resistance, current and voltage are out of phase by a quarter of a cycle and over any period the power consumed averages to zero; the generator is frustrated. In between these extremes the load can be regarded as a mixture of resistive load and a reactive load. A strict watch is kept on industrial consumers to make sure that their plant behaves like domestic appliances in order to correspond to the output of the generating plant. The generating companies can handle a modest reactive load provided it remains constant (their own transmission lines introduce some), but there is strong objection to changes in the reactive load, because they cause a voltage drop on the whole network.

The SPS magnets appear to the power source as a load of varying resistance which requires a varying d.c. voltage to be applied across it. To produce the d.c. directly from an a.c. input, rectifiers must be interposed which are fed through transformers. These transformers look electrically very like magnets with coils tapped at the mid-point, so that current flows alternately through the two halves. But, as with a magnet, when current flow from one half ceases, as the rectifier it is serving cuts off, a reverse voltage is induced which the other half must surmount before its own rectifier can start to pass current. Each time the mains voltage reverses, there is a pause before current starts to flow again. Moreover, in order to control the d.c. output voltage from the system (the rectifiers are followed by capacitors which act as a sort of reservoir that is topped up by the rectifiers) the time at which the rectifiers are made to come in on each half cycle is varied. Again, there is a distortion of the a.c. waveform, which is similar to that created by electronic dimmers on lights, where the average current flow is varied by modifying the fraction of each half cycle during which the lamp is allowed to take current. This may be economical for the consumer, as no power is wasted in heating an alternative (dark) load, but such distortions of the waveform appear in the mains as a reactive load plus a range of harmonics, i.e. frequencies higher than the 50 Hz. The latter can be filtered out, and a constant reactive load can be com-

7/Power and Water

pensated by introducing another phase shift in the opposite direction, for example, by putting across the input a capacitor bank; but a continuously varying reactive load poses an altogether different problem. If a few hundred thousand consumers were regularly in synchronism to turn up and down their electronic dimmers, the effect on the mains voltage would be catastrophic.

In the past, it was customary to get over all the problems of matching accelerator power demand to the supply by interposing between the mains input and the synchrotron magnets a motor–generator set. The mains supplies power continuously to a motor with an extended shaft, on which a heavy flywheel is mounted. The same shaft forms part of the rotor of a generator, which becomes a motor in its turn when the magnets are returning power to the system. The main motor slows down a little as the magnets are energized, and speeds up again over the rest of the cycle. The variation in load demand is reduced to an acceptable level and the varying reactive load is seen only in the machine circuits. However, one of the precepts of engineering is that rotating machines need a lot of maintenance (what car owner would disagree?) and accelerator power supplies more than most. The reason for this is that the torque in the drive shaft of the motor–generator set reverses twice per cycle as the generator turns into a motor and back again. No material likes that kind of treatment. If you are near the power supply of the CERN PS, there is no need to go to the control room to find out the machine cycle time; the whole floor on which the generator stands lurches perceptibly at the moment when energy starts to flow out of the magnets. Not all accelerator rotating power supplies have shaken themselves to pieces, but all have given trouble at some time. The PS power supply is already quite big enough, and for the SPS there was a great incentive to try another route.

Regularly varying loads are not unknown in industry. In a rolling mill, as the head of each slab is fed into the mill, the power demand rises sharply, falling again equally sharply when the tail has gone through. However, techniques have been developed which make it possible to drive such mills from the public supply, provided there is a direct link to a major grid network which can distribute the load. But even the biggest rolling mills have a peak demand of only half that of the SPS, and there is no stored energy to be dissipated because the mills keep turning; the motors are not braked to a stop after each pass.

Undeterred, John Fox, at the Rutherford Laboratory, decided to turn his attention to devising schemes which would allow accelerators also to be powered from the grid, using static compensators to sort out the basic difficulty of matching the current to the voltage, leaving a varying resistive load to be handled by the generating stations. To test out these ideas, the Nimrod synchrotron at Rutherford in 1968 was hooked into the grid during high load and low load periods, and while the accelerator took pulses of 70 megawatts, the voltage and frequency swings on the Electricity Generating Board's network were measured at various points. The swings proved to be lower than expected, and only by careful analysis of the noise on the mains could the Nimrod contribution be picked out. Encouraged by the experiment, CERN made similar tests on the much smaller supply system of Geneva, and in the light of the results decided to equip the PS booster, then nearing the end of the design phase, with a static supply pulsing at 6·6 megawatts peak.

When the Working Group on the SPS power supply was set up to consider in particular how the main ring magnets could be supplied and controlled with the necessary precision, it brought together the experience of many laboratories, but formally it consisted of just two people—John Fox and Simon Van der Meer—complementary in skills and quite opposite in temperament. Fox was an expert on heavy electrical plant, originator of the static compensator idea, able and ready to do battle with the best in the heavy electrical industry on any question relating to big and varying loads. Van der Meer, a physical engineer from Delft, had spent four years with Philips at Eindhoven working on high voltage equipment, electron microscopes, and electronics, before joining CERN in 1956, where he was concerned with the PS power supplies, and such diverse subjects as beam optics, scanning tables for bubble chamber films, and the focusing system of the neutrino beam. He was made responsible for the muon storage ring magnets, where precise control of the magnetic field was of the essence, and then the power supplies for the ISR, where the problem again was one of precise control rather than novelty of supply. Immensely conscientious and reserved to the point of being withdrawn, his staff who were recruited when he was made head of the SPS Power Supplies Group took time to understand that his rare smiles give the real clue to his deep reserves of charm and sincerity. Charm and reserve are not the words which come to mind when thinking of John Fox. Outspoken to the point of being aggressive, he is never so happy as when beating the system of whatever officialdom stands in his way. But he can, if in the mood, hold an audience spellbound with a continuous flow of brilliant anecdotes and shrewd comment, spiced by a devastating wit. When he later joined Lévy-Mandel's Group to look after the heavy electrical site installations, it said much for CERN's anationalism that even his particular brand of untrammelled chauvinism could be accepted without incident.

Fundamental to the question of whether a static power supply could be built for the SPS was the ability of the network in the region to absorb the overall pulsed load. It was clear that the West-Swiss network could not cope, and the agreements

7/Power and Water

Peak power consumption of the SPS is comparable to the average consumption of Geneva. To supply this a special grid line is brought from a power station at Génissiat, some 35 km down the Rhône, which in turn is connected to the European grid network. The substation terminating this line at the SPS is the largest substation serving a single customer on the system of Electricité de France. Apart from the size of the load, the surges in power caused by the energizing of the main ring magnets, followed by a return of the stored energy to the system, necessitate a direct link with a big network if the six-second rhythm of the SPS is not to be felt by other consumers in the region.

Two transformers convert the 380 kV input to the 18 kV supply distributed round the site. One of these is reserved for the main ring power supplies, while the other feeds power to the rest of the site.

covering the project authorization stipulated that electrical power would be supplied by France. Discussions were started with Electricité de France (EDF). It was not easy to find a load of the size envisaged with which to perform experiments, but, with the collaboration of the French and British authorities, the cross-channel supply link was pulsed, and measurements made of the effects on the continental grid. These provided data for the design of a system which would be coupled to the grid at Génissiat, some 35 kilometres down the Rhône from CERN. Fortunately, Génissiat is not only a hydro-electric power station of EDF, but also a cross-roads for the continental grid network, and the surges could be shared amongst a number of generating stations.

A special 380 kilovolt line was erected from Génissiat to the SPS site by EDF at CERN's expense, an arrangement that took the delegates of some member states a

7/Power and Water

In addition to the distribution boards, to be expected in a sub-station of this size, is an unusually complicated installation for the measurement of the power taken and returned to the system. To calculate the power bill it is not sufficient to know only the average consumption, but also the detailed pattern of power flow; in addition, the way in which the current flow varies throughout the 50 Hz alternating cycle—and to what extent it is out of phase with the line voltage. A constant discrepancy is not serious, but a varying discrepancy upsets the line voltage, and is reflected back throughout the system.

little by surprise, as in the early enquiries for site proposals it had been laid down that power would be brought to the site at the expense of the host state. However, circumstances had changed, and it was recognized that France had not chosen St. Genis-Pouilly for the SPS site, and that the grid line was needed solely for the project and would serve no-one else. Detailed specifications were drawn up for the shape of the power curve, to minimize the effect of the power surges. In particular, it was agreed that the cycle time of the SPS would never be less than three seconds, there would be a 60 millisecond rounding off between the accelerating phase and the steady state, and a further 120 millisecond rounding off after the protons were ejected, and before power was returned to the system. The whole cycle was optimized to minimize the total cost of the power supply, plus the capitalized cost of power consumption. These parameters established the peak power that could be taken by the magnets.

Fixing the power cost called for a lot of hard bargaining. The extreme position that CERN could take was that it paid simply for the mean power consumed, averaged over a long period; EDF on the other hand could maintain that it was the peak power that determined its supply commitments, and any return of power was just an added nuisance. In the end, a complicated scale was worked out which added to the mean consumption charge a charge for the peak power used, taking into account the expected peak power demand announced in advance, plus a charge for the complexity of the cycle and the residual reactive load. How to measure these numbers had also to be agreed, and the monitoring system necessary is much more elaborate than the familiar meter in the cellar.

When industry was asked to tender for the compensating system, a number of possibilities were left open, as there are several ways in which, in principle, the power factor can be stabilized, and then compensated. All are highly complicated, because not only is the required voltage–current pattern in the magnets complex, but also there is a three-phase input supply. Finally, the system accepted was one offered by a British company based on the use of a saturable reactor. This has nothing to do with nuclear reactors, but is a type of complex multi-legged transformer in which the three mains phases interact with each other, and where advantage is taken of the magnetic saturation effects in the iron core. Coupled in series with the supply to the rectifier stations, the reactor introduces a power factor of the same sign as that of the rectifier stations, but one which gets smaller as the current rises and the magnetic field saturates. The rectifier stations have the opposite characteristic, and so the total reactive load remains substantially constant. Across the input is a capacitor bank which makes the necessary compensation to convert the load to one that is essentially resistive. The net result is that when the SPS is operating at full power the voltage swing at Génissiat is only about 0·3 per cent. Two 90 MW transformers are coupled to the grid line at CERN to transform the line voltage down to 18 kilovolts and space has been left for a third. One transformer is reserved for the magnet power supplies, and the other for all the other supplies to the site. This effectively isolates the rest of the machine from the effects of the power surges.

Sorting out the mains input is, however, just the beginning of the power supplies problem. The magnets have to be supplied with a d.c. voltage which

At each of the six auxiliary buildings, two groups of four transformers feed the rectifier stations, which convert the a.c. input into a varying d.c. supply and energize the bending magnets in the main ring. Filters take out the mains ripple that is left. As all the bending magnets are connected in series, just one power supply can feed the whole ring, but with only $\frac{1}{12}$ of the current that can flow when all twelve are operating. To cut down stray magnetic fields created by heavy currents through the connecting cables in the tunnel, alternate groups of four magnets are connected through one cable run, and the others through a return cable. The focusing and de-focusing quadrupoles are each fed by a separate system, but the precaution is taken to run the cables through the tunnel in opposite directions.

varies between practically zero and an integrated total of nearly 25 000 volts and the current has to change from almost nothing to a maximum of 5000 amperes all with a precision of better than 1 part in 10 000. In addition, in order not to interfere with the slow extraction system, the ripple on the supply has to be less than 1 volt from each individual supply station. Even the position of the cable runs in the tunnel matters, as it is necessary to eliminate stray magnetic fields which could upset the beam guidance and focusing.

Twelve series-coupled rectifier stations have been installed, two in each of the auxiliary buildings, feeding alternate groups of four bending magnets in the adjacent sectors. The current is led round the ring through one station in each building in turn, and back through the others in the opposite direction. Two further rectifier

Simon Van der Meer, head of the minuscule Power Supplies Group, which was concerned not merely with the supplies for the main ring magnets, but also the design of hundreds of auxiliary supplies. His genius in devising the control system of the main power supplies through a single computer, and the intricate feed-back loops within the system, has given the SPS extraordinary precision and reproducibility. At the same time, the disturbance to the grid voltage is minimal. The computer program he developed contains thousands of instructions.

stations located in auxiliary building 3 (BA 3) feed the two sets of quadrupoles independently and in opposite directions round the ring. Nothing is tied to earth, so the system can find its own mean level. Each rectifier station consists of four rectifier bridges connected in series and 'fired' sequentially. The 600 cycle per second ripple so produced is smoothed out by chokes and capacitors, whilst a dynamic filter, introducing a nullifying ripple in the opposite direction, can be brought into operation if additional smoothing is required. At the heart of each bridge is a pair of thyristors on each phase, made of single crystals of silicon that

7/Power and Water

become conducting when a subsidiary voltage pulse is applied. They shut off again when the alternating voltage on the input reverses. Playing tunes on the sequential firing of the 24 thyristors is how the d.c. supply from each station is varied between zero and 2100 volts. Subtle combinations of the output from each station, allow the current in the magnets to be altered smoothly, whilst avoiding sudden changes in the reactive load on the mains. Although separate calls for tender were sent out to industry for the 18 kilovolt transformers, the rectifier stations, and the filters, it was a single German company that made the best offer for all three.

Behind the massive elements in the system is an impressive array of what is called low-level electronics, the adjective referring to the power consumed and not the level of ingenuity needed in designing the control circuitry. The decision work is carried in a dedicated computer, which prepares the coded tables from which the voltage output required from each of the rectifier stations is read off every 10 milliseconds (see Chapter 10). From the signals received, a waveform generator at each station develops a facsimile of the voltage pattern, then circuits internal to the station measure the difference between the output registered and the reference, and automatically adjust the time at which the thyristors fire to give a null reading. A nice balance is needed to obtain a rapid response to the changes demanded and drifts in the input voltage, without creating a situation whereby the system 'chases its own tail' and begins to lead an independent existence.

Quite apart from the main magnet power supplies and the special power supplies for such machine components as the injection and ejection magnets designed by the Group concerned, the Power Supplies Group was required to devise, order, and commission nearly 200 auxiliary power supplies with their control circuits. A certain amount of standardization was possible, but over forty different types were involved. From the manufacturing side, the advantages of standardization are not great; rectifiers are an off-the-shelf item, but transformers are wound to order, and it is more economic to define for each supply such things as maximum current and voltage outputs, whether a positive and/or negative output is needed, how much ripple can be tolerated, etc., than to try and produce a general purpose design. It was convenient, nevertheless, to go out to tender with a bulk order. A Dutch–British consortium was the successful tenderer.

Power supplies are not, perhaps, the most exciting of the machine constituents, but their stability and reliability are at the bottom of every aspect of the machine's performance; without them, nothing else can function. In the constructional programme they were among the first items, after the buildings, that had to be ready, so that other equipment could be developed and tested. They are also a continuing concern. The accelerator is far from being the only consumer of electrical power.

In the rectifier stations, the output voltage is controlled by varying the moment at which the rectifiers fire, in a manner comparable to an electronic light dimmer which shuts off part of the mains cycle. Internal feed-back circuits measure the output voltage, and compare it with the computer's instruction for that moment, while a continual check is made of the magnetic fields actually being produced, so that the computer can compare the performance of the machine with the operating instructions.

Every beam line requires its own special supplies for guidance and focusing magnets and analysers, and every experiment has its own special needs. Electricity consumption in the experimental zones will exceed the mean power consumption of the accelerator, which is much less than the peak demand being in the region of 50 megawatts. When all the big detectors are in operation, the West area alone will need up to 30 megawatts of electricity, and will strain to the limit the present supply capacity of the sub-station connected to the Swiss network which feeds this area. It will probably be necessary to programme the research so as not to exceed the maximum power available to the original CERN site of 80 megawatts.

The chopping of the mains input wave, as well as the interaction between the transformers and rectifiers, results in a changing lag between current and voltage in the mains, as the current taken by the magnets alters. To prevent this affecting the grid system, the mains are fed through a special transformer-like device whose lag changes in the opposite direction, so keeping the shift constant. Finally, this shift is compensated by a battery of capacitors across the input.

Power demands in the North area will grow as the experimental programme develops, and must be expected to outgrow that of the West area. In addition, there are the normal laboratory supplies to take into account. All these extra demands require a very considerable installation of 'conventional services' around the site. These came into the extensive province of Lévy-Mandel, as did the other service which figured early in the Site Installation's planning schedule: that of water.

It was agreed at the beginning of the project that the supply of cooling water would be the responsibility of Switzerland. The Federal authorities commissioned the cantonal public utility to carry out the initial survey, but stepped in again with their own expert when the principles of the scheme were all but finalized. He promptly began a new study, and came up with counter proposals which meant that negotiations had to start all over again. At one time it looked as if the whole building programme would be delayed because of the absence of adequate cooling water supplies. A provisional supply was laid on from the existing CERN mains for development work and laboratory use, but this could not cope with the demand once serious commissioning began. The main point at issue was how to dispose of the hot reject water from the primary cooling system, as the Confederation was unhappy at the idea of rejecting it by any means directly into the Rhône. The law requires that the rise in temperature created by an outflow into a river should not exceed 3° C, but the law is somewhat vague in defining where this temperature should be measured, and what degree of mixing should be considered reasonable. A further element in the discussion was the plan to build a power station near the proposed outflow and, although in comparison with the reject heat from this source, CERN's contribution would be small (of the order of 5 per cent), the Swiss authorities had become very sensitive to public reaction to the thermal pollution of the rivers. There was, also, an international aspect as the Rhône becomes French just a few kilometres downstream.

Finally, it was decided that CERN could reject its water into the Rhône via a stream, the Nant d'Avril, which runs between the site and the nearby village of Meyrin, provided the maximum temperature at the CERN outlet did not exceed 23·8° C, and the rate of change of temperature did not rise above 1° C per 90 minutes.

There was little disagreement about the intake from Lake Geneva, although the CERN sailing club was instructed by the local police to remove the buoy marking its position on the assumption that it was a stray regatta mark. Once this was cleared up, the buoy was replaced some 1200 metres out in the lake at Vengeron, half-way between Geneva and Versoix, at a point where the water is about 40 metres deep. Water is taken from the bottom of the lake through a 1-metre pipe, made up from long sections welded together on a convenient quay on the opposite bank, floated across and sunk. It is pumped through a purification plant on shore, and through a 1-m conduit up to two 5000 m³ reservoirs situated near BA6. The reservoirs are concrete tanks with a domed roof, built partly underground and covered

Water taken from the bottom of Lac Léman is piped through a purification and pumping station into two 5000 cubic metre reservoirs, photographed when the one on the right was complete, and the one to the left still under construction. The water from the lake is not used to cool the machine components directly, but, to avoid any possibility of radioactivation, is used to cool the water flowing in a closed system serving only the SPS. The return water is cooled (or, until start-up, heated) before being rejected through a pipeline to the Nant d'Avril, a small watercourse which finds its way eventually into the Rhône. Strict control is kept not only of the maximum temperature, but also of rates of change of temperature.

with earth so that a grassy mound is all that is visible from the road. On its way, the conduit passes near an existing water source, and as it was clear that the complete installation could not be ready early enough, arrangements were made for the reservoirs to be filled from this source to allow commissioning to go on unhindered. This same connexion can also be used in emergency if the Vengeron station breaks down. As the purification requirements for the cooling plant and conduits fell little short of drinking water standards, and the supply line was to be coupled to a source of drinking water, the decision was taken to install at Vengeron a treatment plant able to deliver water suitable for drinking purposes. This meant that no additional plant was needed on the site for 'domestic' uses. It also allowed the local commune to arrange, at an advantageous price, a supplementary feed to its own system, and one-third of the water taken from the lake, even during full operation, is destined for the new apartment blocks in Meyrin. One of the most difficult problems was negotiating the many way-leaves through the local farming land along the route of the conduit, and some changes of plan were needed to ensure that the first link-ups could be finished within the revised programme schedule.

The rate of heat dissipation required is roughly equal to the electrical power converted, and corresponds to a maximum flow of 700 litres per second, to which must be added a flow of 300 litres per second for general use, making a grand total of 1000 litres per second: this is equivalent to a fast-flowing stream 2 metres wide by 30 centimetres deep. Cooling water is pumped round the machine to each auxiliary building in turn through a pipe buried a metre or so underground. The section of the pipe diminishes after each take-off point, while alongside the parallel return conduit progressively increases in section. In the same trench as that dug for the water supplies are laid the electric cables linking the auxiliary buildings to the control room. A small reservoir situated at the highest point on the site, between BA3 and BA4, acts as header in the return system, and the flow is adjusted to maintain the level in this reservoir constant. Leading away from the main ring mains near BA3 are two supply feeds to the North experimental zone, one from the cold and one from the hot leg, with a single return into the ongoing hot line. Most of the machine components, and all those installed in the tunnels or elsewhere if there is any radioactivity, are cooled by demineralized water circulating in a closed system. This secondary system gives up its heat, in heat exchangers mounted in the auxiliary buildings and experimental areas, to the water flowing through the primary system.

Four forced-draught cooling towers erected close to the reservoirs and pumping station reduce the temperature of the reject water to an acceptable value. As the

7/Power and Water

The water is piped round the site in two conduits; one for the cold supply, which diminishes progressively in section as it proceeds round the ring, and one for the hot return, which increases in section. A small header reservoir at the highest point of the site maintains the pressure constant.

inlet water temperature varies between a minimum of 6° C in winter and a maximum of 13° C in summer, and as the temperature rise in the heat exchangers is limited to 15° C, the maximum temperature of the return flow should not exceed 28° C. Even so, the cooling towers have been designed to be able to meet the reject water temperature limit of 23·8° C, with a full flow in of 31·5° C, and a wet bulb air temperature of the high summer norm of 20·5° C. In summer, they can also be employed as pre-heaters of the water in the return conduit by circulating the return water through them in a closed loop. If there were no pre-heating, rapid start-up of the accelerator after an extended shut-down would not be possible without exceeding the maximum rate of temperature rise permitted for the reject water. It takes about 100 minutes for water fed to the first heat exchangers in BA6 to make the tour of the machine, and emerge at the conduit outlet. Taking into account the losses through the walls of the conduit into the ground, the outlet temperature will settle down to a stable value after about 120 minutes. If, then, no precautions were instituted, the temperature of the water in the return conduit would initially be the same as that of the inlet water, say 13° C, and there would be a rise of 15° C over the first 120 minutes following machine start-up. The corresponding rise in the temperature of the reject water would be from 13° C to 23·8° C. Even allowing for the additional hold-up time in the cooling tower ponds, the rate of rise would far exceed the limits laid down. To get over this difficulty, hot water is re-circulated through the return conduit before start-up. If the shutdown has been short, re-cycling may be all that is needed, but after a long shutdown, re-heating is necessary. In summer, at least, this can be done partially by re-circulation through the cooling towers. For rapid start-up an initial temperature of 20° C is, however, still too low, and heaters have to be employed. In winter, the cooling towers can give no help, and the heaters must assume the whole load. Protecting the local fish is not a cheap affair.

8/Ins and Outs

A nice mixture of elegant steering and brutal kicking is needed to feed the protons into the SPS at the right time and with the right distribution and then, once accelerated, to get them away to the experimental areas as required.

In the PS they are wheeling round the 630 metre circumference in 20 bunches, and by the time their energy has been raised to 10 GeV they are focused into a stream of pulsating packets with an average cross-section of about 2 cm². They make a complete turn of the PS ring in a little over two-millionths of a second, and the clear length between bunches is roughly twice the length of an individual bunch.

Techniques for kicking them out one bunch at a time were well known, but if they were introduced into the SPS as separate bunches there would be a need to wait, before starting the next phase of acceleration, until each bunch had drifted into a ribbon 350 metres long. It was better to try to pull them out of the PS as a continuous stream, even if this meant a few were lost in the process. The PS began to operate in 1959, and had not been designed for gymnastics of this sort. Nevertheless, in spite of the elementary control system that was in use, a great deal of experience had been gained in different types of ejection system, and it was felt that a method of continuous transfer should be tried.

The basic idea is to start by letting the bunches spread out inside the PS ring by switching off the r.f. accelerating field. The bunches are formed and maintained by the r.f. field, which is able to keep in synchronism only by giving the slower particles a little more energy than the average as they orbit round, and the faster ones a little less. Removing the r.f. field results in some creeping ahead, and others dropping behind until a more or less continuous ribbon is formed. The quadrupoles in the ring keep them together, laterally and vertically. Next, a stepping bumper magnet is brought into play to push the ribbon over to one side of the vacuum tube, some way up-stream of the ejection point. Those protons swinging out the widest find themselves passing behind a thin foil and into the influence of a strong electrostatic field, established between the foil (the septum) and a high voltage plate. The main proton beam outside the septum, which is earthed, feels no effect, but those behind it are bent away from their original line of flight. The deviation is not enough to get them clear of all the machinery surrounding the vacuum tube; instead they are pushed back in the other direction, and across the path of the main stream. Further on are quadrupoles which exaggerate the angular separation, and bumper magnets which push the whole stream once more over to the outside, when the focusing swing is again in the outward direction. By this time, the separation is sufficient for the detached stream to pass behind the septum of a septum magnet, consisting of a horizontal coil (with the ends bent out of the way) the inner conductor of which is made as thin as possible. Now the deflection can be outwards, and the detached beam can be swung clear of the PS magnets into the transfer channel, where it can be manipulated in comparative comfort.

The main stream sees no magnetic field on the other side of the septum, and continues to circulate round the ring. By the time those left inside have made a complete turn, the ribbon has assumed a different form, the Q-value being so adjusted as to bring a different group of protons into the outside position when arriving for the second time at the first bumper. This is programmed to give now a stronger bump, so the whole process can be repeated. The art is to change the energizing currents in the bumpers every two-millionths of a second, to move the beam over just the right amount, and to set the Q-value so that a constant fraction of the beam is scraped off on every turn. By the middle of 1972, using equipment for the most part originally built for other purposes, it had been proved possible to extract over 95 per cent of the beam circulating in the PS with remarkable uniformity over exactly eleven turns. Moreover, the ejected stream was significantly thinner than the original bunches, in consequence would be easier to handle in the SPS, and would take up less space in the vacuum tube there. Eleven turns of the PS is equal to the circumference of the SPS, and the new machine could be designed for continuous transfer of just the right duration, with the confidence that when the time came, the PS would be ready to supply what was needed.

It is worth pausing for a moment to reflect on the lethal properties of a beam of elementary particles accelerated to high energies. Intense beams of electrons of much lower energy are regularly used in industry for welding and cutting, the energy being chosen to match the depth of heating required. At 10 GeV the penetration of a beam of protons is already several centimetres of steel, and in slowing them down, a great shower of energetic particles and gamma rays is produced which bombard the surrounding region in a narrow cone extending over metres. At 400 GeV the distances are correspondingly bigger. In a septum magnet some of the beam is inevitably going to collide with the septum, and if the number is too great the septum will have a very limited life, and the surrounding apparatus will

8/Ins and Outs

Bas de Raad, as head of the Beam Transfer Group, was responsible for guiding the beam from the PS into the SPS ring, and, when accelerated, extracting it in whatever manner was required and leading it to the experimental areas. This involved the design not only of the guiding and separating devices, but also their power supplies, some of which would be required to generate hundreds of thousands of amps for closely defined periods of time. When the machine was ready for commissioning, he was put in charge of the Running-In Committee, and continues now as the man responsible for the complete accelerator.

quickly become too radioactive to handle and insulation will suffer. With this in mind, the electrostatic septum in the PS was designed for easy and rapid replacement, but in ejecting from the SPS the beams would be 15 times as stiff and unbendable, and a clean separation of the fraction to be extracted proportionately more difficult.

Bas de Raad (no-one ever uses his full first name of Bastiaan) was the man chosen to steer the protons across to the SPS and into the vacuum tube, and to cope with the more tricky problem of feeding them out to the experimentalists when their energy had been raised to 400 GeV. He had previously been responsible for the delicate job of nudging the beams into the ISR, and his methodical and meticulous temperament made him ideal for the task of designing the special devices needed for the SPS. A Dutchman, at CERN since leaving University in 1954, his big and ungainly appearance disguises a stickler for detail, a man never satisfied until he is doubly sure that what he has proposed will fulfil the specifications laid down.

The techniques for feeding the beam from the PS into the SPS were well known, but the margins for error were exceptionally tight due to the small cross-section of the SPS vacuum tube, and the length of time over which constant conditions had to be maintained during the actual injection. The stream from the PS travels along the ISR feed tunnel for a distance of 300 metres and then branches off, dipping down under the Geneva–St. Genis road to arrive at the SPS ring 800 metres farther on and 40 metres underground. All along the line is a regular pattern of quadrupoles (made in the Netherlands) while bending magnets, originally built for a now abandoned link between the ISR and the West experimental area, take care of the curves. Coming into the SPS, the beam shaves past one of the focusing magnets there and then passes through two d.c. septum magnets, following which it enters an enlarged section of the main ring vacuum tube, up-stream of the next quadrupole. It is now approaching the straight-through line at an angle of $\frac{1}{3}°$. Two bumper magnets on the far side, of the window-frame type with cores of ferrite (the magnetic material used in tape-recorder tapes) pulsed for a time equal to one revolution in the SPS ring, swing the beam into its proper orbit.

Once the ribbon is orbiting comfortably, the r.f. system shuffles it into bunches—4620 of them, distributed regularly round the ring, travelling at almost the speed of light, spaced at a distance of 1·5 metres between centres, after which the bunches can be accelerated up to the energy required. This will depend on the particular experiments going on. Not everyone wants protons of the maximum energy, nor a supply spread over the same length of time. The bubble chamber physicists demand a short sharp burst, the physicists working with electronic detectors a slow, steady flow, and those working with bubble chambers and electronic detectors in tandem something in between. The ejection system has to be able to satisfy them all, and several users on every pulse.

Fast ejection cannot be done by kicking out individual bunches. Apart from the fact that the time between bunches is only one two-hundredth of a millionth of a second, which does not leave time to do very much, the holes that would then be punched out of the circulating beam would upset the r.f. system, and it would be very difficult to handle what was left. The alternative is to scrape off the required

8/Ins and Outs

Injection of 10 GeV protons might be considered now to involve relatively standard techniques, but extracting a beam of 400 GeV in a controlled way is a different order of challenge. At some point, the beam of protons must be pushed into such a position that a pre-determined fraction comes under the influence of an electrostatic or magnetic field which the rest of the beam does not feel. In this electrostatic septum, the main beam passes through the aperture, above and left of centre, while the chosen fraction passes to the right of a thin knife edge, or septum. A strong electric field is generated between it and the polished plate carried by insulators from the walls of the surrounding vacuum vessel and to which a tension of about 200 000 volts is applied. The deviation imparted to the protons is just the beginning of a series of steps to eject them from the machine.

fraction, much as is done in the PS when feeding the beam to the SPS. But now the beam is 15 times as stiff, and the effort needed to cut part away and the strain on the 'knife' are all that much bigger. There is, however, room to manoeuvre, and if a septum could be devised which would stand up to the punishment it would

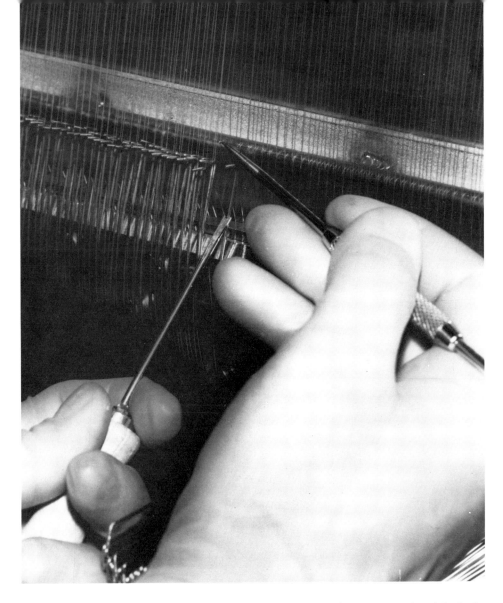

To avoid distortion of the knife edge, as would occur if it were made of foil, the septum is made up of fine wires stretched across a strong supporting fork by individual springs, which flick a wire out of the path of the beam if it should be burnt through. The wires are of tungsten 0·15 mm diameter, spaced 1·5 mm apart, and over 2000 are used in a unit to give a deflection of one eightieth of a degree.

inevitably get, the separated beam could be led clear of the ring machinery without much trouble.

    A foil would not be appropriate. It would be practically impossible to make one perfect plane over the length needed, and even if this could be done, the heating of the leading edge which would take the brunt of the damage would cause distortion and quickly aggravate the situation. Moreover, once damaged, the complete septum would have to be replaced, a lengthy and costly process. An ingenious solution was to use instead of a foil, a row of fine wires stretched across two rigid supports, each wire being tensioned by a spring, top and bottom. Should one of

A line of electrostatic septa mounted in the ring (viewed from downstream with reference to the direction of the proton beam). Two identical extraction systems have been installed, one in straight section 6 to supply the West Area, and the other in straight section 2 to supply the North Area with either a sudden pulse, a long drawn out feed, or some compromise in between.

the wires break, the springs will flick the severed ends out of the danger area, and the loss of an odd wire, here and there, makes little difference to the surrounding field characteristics. Some 2080 tungsten wires, 0·12 mm diameter spaced at 1·5 mm intervals, go into a single septum unit, four of which are needed to bend the separated beam sufficiently for it to be guided into the septum magnets. The wires are earthed and the other electrode, of highly polished anodized alloy, 3 m long, is maintained at −200 kilovolts. The two electrodes must be parallel to an accuracy of $\frac{1}{20}$ mm, whilst the gap between them is adjustable from 10 to 30 mm. Final alignment is carried out *in situ* using infrared detectors to measure

Following the bank of electrostatic septa are four pairs of septum magnets with septa of 4 mm, and further on a group of five pairs of thick septum magnets with a separating conductor of 16 mm. The main beam passes through the body of the surrounding vacuum chamber, while the extracted beam is channelled through the window in the steel core of the magnet.

*A close-up of the end of the septum magnet shows the group of conductors (**left**) which divide the high field area in the magnet aperture from the free field area outside. Around the coil formed by this conductor and the inner conductor, a current of up to 24 000 ampere flows when the extraction system is in operation.*

*The final extraction stage in the tunnel, where from the end of the last group of magnets the ejected beam line can be seen (**right**) separated from the tube in which the beam circulates round the machine. From this point on, with the ejected beam clear of the machine components, its focusing and guiding becomes a relatively straightforward matter.*

Downstream of the extraction point, when the ejected beam is already some metres away, to the right, are further pulsed magnets, brought into use when part of the beam is ejected to correct the 'bump' left in the circulating beam. As with all these magnets, from the outside all that is visible is the vacuum tank enclosing the assembly.

the temperature rise in the wires. A surprising bonus was the high electric gradient the system would support, as theory had predicted that flash-over would occur at lower gradients than when using foils. In fact, it is the reverse.

Following the electrostatic deflectors, 50 metres downstream, are four pairs of septum magnets with septa 4 mm thick, carrying currents of up to 7500 A, and further on again, five pairs of septum magnets with 16 mm septa carrying currents of up to 24 000 A. All these separating units are contained in stainless steel tanks pumped down to low pressure. Two identical systems are installed in straight sections 2 and 6, feeding respectively the North and West experimental zones.

Slow and smooth ejection over tenths of a second cannot be done in the same manner as fast ejection, as the precision required in matching the strength of the bumping magnets to the changing form of the residual beam would be impossible to achieve. Here, however, advantage can be taken of the tendency of the circulating beam to run into resonances, and the dependence of Q-value on the proton energy. When the desired mean energy has been reached, the current in the main quadrupoles is adjusted to bring the Q-value close to a resonance. Sextupole magnets at a point upstream of the ejection channel are then programmed to sweep the Q-value locally through the resonance. Successively, protons falling within a narrow energy range and having orbits of a particular form will be caught and will swing out wider and wider on successive turns, finally passing behind the septum of the first electrostatic septum unit and on down the extraction channel. For a given current in the sextupoles, only those protons with the right energy and orbit have their swings amplified on every turn, the others feel a temporary effect but quickly fall out of step. By slowly changing the sextupole power, all the protons in the main beam can be brought into step in turn and a steady stream can be supplied to the experimentalists.

One other ejection system is needed, to get rid of the beam during machine tuning or when things go wrong. If a beam of maximum intensity and energy is suddenly lost, over one fifty-thousandth of a second the amount of energy being

Opposite

To get rid of the beam in a safe and controlled way, it is deflected by yet another kicker magnet into a block of cast iron with a copper and aluminium core. Power supplies for this magnet are permanently charged so that in the event of a machine failure, or at a signal from the control system, a current of 10 000 amperes over 25 microseconds is discharged, and the circulating beam is diverted into a shielded dump located in straight section 4, well away from all the other special components.

8/*Ins and Outs*

163

Europe's Giant Accelerator

dissipated is about the same as the average electricity consumption in the UK. It is not the sort of thing that can be left to chance. In straight section 4, a dump has been built consisting of blocks of aluminium and copper in cast iron vessels into which the beam can be directed, when necessary, by pulsed magnets. In addition to deflecting the beam out of its normal trajectory, these magnets spread it across the dump to distribute the heat generated. This will become the most radioactive section of the accelerator system, so it has been placed all on its own in that part of the ring where there is 50 metres of rock and soil above.

9/Acceleration

Once safely circulating inside the SPS ring, the protons come under the influence of the main accelerating system. In very broad terms, the principle behind the system is the same as in other synchrotrons or linear accelerators, but in all its detail it is the first of its kind.

A radio wave is set up in a series of resonating cells tuned to the frequency of arrival of the protons which have been grouped into a regular series of discrete bunches. When a bunch exactly in synchronism appears in a cell it is swept forward by the field which in due course will change direction. By then, however, the bunch has passed through a shield into the next cell, where again it finds the field in the right direction.

Meanwhile, the field inside the one it has left grows in the opposite direction, reaches a peak, diminishes once more, is zero for an instant, and then starts to grow again in the forward direction ready to receive the next bunch. In this way, the protons receive a regular series of kicks forward in each cell, steadily gaining in energy all the time under the influence of a series of electric fields whose maximum intensity can be kept within manageable proportions. The protons not quite in synchronism receive a stronger or weaker series of kicks depending on their position relative to the centre of the bunch.

In synchrotrons of lower energy, the speed of the protons rises appreciably with increasing energy, and the frequency of the accelerating field must also rise appreciably. This, in turn, means that the natural resonating frequency of the cells must be continually retuned. The protons make many turns in the machine, so it is not possible, as in a linear accelerator (where they go through once only), to change the lengths of the cells to keep the same transit time, and so operate at constant frequency.

In the SPS, the protons when they are injected at an energy of 10 GeV are already travelling at 99.5 per cent the speed of light, and in going to the full energy of 400 GeV their speed increases by only 0.44 per cent; their increase in energy appears almost entirely as an increase in their effective mass. The frequency of the accelerating field has still to be accurately synchronized to the circulating period,

Clemens Zettler, the longest-serving member of the SPS team, was developing an accelerating system in the early 60s that would have been used in the injector synchrotron had the SPS been built on another site and not been able to use the PS. When the decision was taken to build the ISR first, he continued to work on designs for the future synchrotron and was one of the little group of four which made up the SPS team during the evolutionary phases of the Project, becoming finally head of the r.f. Group. During the brief period of commissioning when the beam was lost soon after acceleration started, he never had doubts that there was something wrong with the calculations he had made or the equipment that he had designed. The suggestion that there were troubles in the r.f. was one of the few things that could make him cross.

By the time the decision on the SPS was taken, the general form of the travelling wave accelerating system was already well advanced. In its essential, it consisted of a line of resonating cells separated by drift tubes, through which the beam of protons passes, in which an oscillating electric field is made to resonate so that there is an accelerating voltage along the line of the beam. The bunches of protons when they emerge from a tube, experience an accelerating field which changes as they go through, so that those with more energy than the mean value are accelerated less, and those with less energy are accelerated more. As a result, the bunches are automatically kept together, the front and back markers continually changing places. Around the ring, are 4620 bunches spaced at 1·5 metre intervals, while the distance between the centres of the drift tubes is one quarter of this value.

but it has been found possible to devise a system where the cells do not have to be continuously retuned, and can be fixed once and for all. Moreover, the efficiency with which input power is converted into proton energy is reasonable, and increases with increasing beam intensity.

The basic principles of this new approach to accelerating systems were first worked out by Wolfgang Schnell, a quiet, modest, German physicist who had shown his genius when building the accelerating system of the PS. He became part of the original study group of the 300 GeV project, but moved over to take charge of the subtle system of stacking protons in the ISR when approval for this project was given at the end of 1965. Clemens Zettler was then working on a different system for a fast cycling synchrotron which was to be the injector for the big machine had it been built elsewhere. He stepped in to take charge of the studies, and in due course turned the ideas of Schnell into an engineering complex which would match the particular design constraints of the SPS.

The system is called a travelling wave accelerator, and it is tempting to liken the behaviour of the proton bunches to a surf-rider who is picked up on the crest of a wave and swept to the shore riding continuously down the wave slope. The parallel is, however, far from exact, and we must look for other analogies to understand the mechanism.

Let us begin with the physical structure of the two accelerating cavities, located in straight section 3: from the outside they give the appearance of high-pressure boilers rather than of boxes of electronic equipment. Each cavity, built up from five units, is 20 m long and 75 cm diameter. The shell and covers are made of copper-clad steel. Inside are 55 evenly spaced copper 'drift tubes', cylinders 15 cm long with a 13 cm bore, centrally supported in the cavity by two horizontal hollow stems, through which flows cooling water. The tank itself is also cooled by passing water through a series of coils glued to the wall.

A single cell is defined by the volume enclosed by the shell and the drift tubes with their supporting stems on each side. This volume resonates at the chosen operating frequency of the system (leaving aside the small swings that occur during each accelerating cycle) of 200 megahertz, a frequency typical of those used in television channels. The drift tubes not only define the cell size, but also with their stems determine how the electric radio wave when fed in at one end will flow down the cavity. They have a delaying effect, rather as if the wave had to run up and down the stems before deciding to go on to the next cell. The timing is such, however, that when the wave arrives in the next cell it is just one quarter of a wavelength different from the one before. Imagine a string of flag wavers lining the route of a procession. At a given signal, the first one begins to wave at a fixed frequency. After a certain time, the next one in line begins to wave at the same frequency, choosing the moment to start when his neighbour has reached the end of his swing. In due course, the next one in line reacts similarly, and so on down the line. Once the whole line is in motion the delays in each of them starting will no longer be evident, and it will appear as if a wave is going along the line. Moreover, depending on whether at the start each flag waver decided to start to wave in the

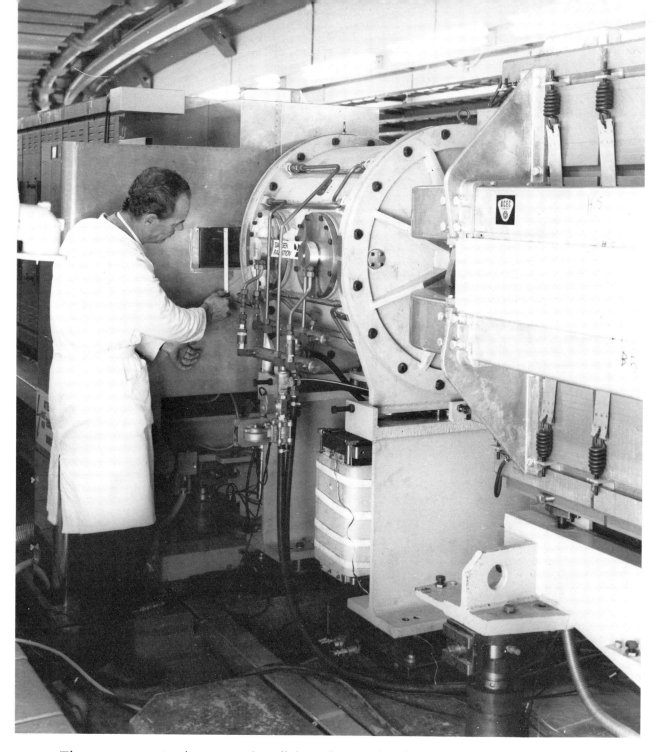

The exact resonating frequency of a cell depends upon the physical dimensions of the body and the internal components. A two-cell cavity was built to confirm the theoretical studies, and was inserted in the ring of the PS to check out the interaction between the cells and a circulating beam of protons. This form of cavity has the characteristic that as the intensity of the beam increases, so the efficiency of the transfer of energy goes up, a desirable characteristic for a machine that will certainly be upgraded in intensity as the years go by.

Two cavities are employed, each of which is made up of five sections containing 11 cells. The form of the drift tubes and supporting stems is such that as the wave travels along the tube it is delayed for a period between each, a period adjusted to make each bunch of protons experience the same electric field in each cell. In effect, the bunches of protons are riding down the wave in each cell, and every fourth cell is 'ringing' in tune. The distance between the centres of these cells corresponds to the distance between bunches, and the wavelength of the r.f. oscillation.

9/Acceleration

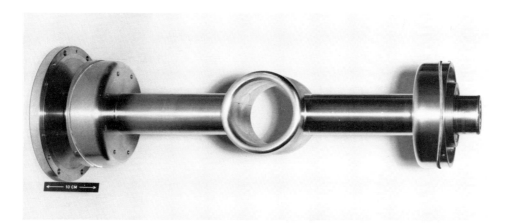

Final tuning of the cavities is made by adjusting the diameter and length of the stems supporting the drift tubes. When the cavity sections were received from the manufacturers, a standard set of 11 stems and drift tubes was mounted in position, and the resonant characteristics of the section measured. From these measurements was calculated the dimensions of the stems necessary to tune the cavity to the desired resonant frequency of about 200 MHz. A special procedure was developed in the CERN workshops then to braze the machined stems to the end fixings with the precision necessary.

direction his predecessor had been finishing, or in the one he was about to begin, the wave would seem to go up or down the line.

For reasons of efficiency, the power is fed into the accelerating cavities at the downstream end and moves upstream, but the wave is made to travel in the opposite direction, which is the direction of movement of the protons. It takes a few moments (about a millionth of a second, which is quite long when compared with the oscillation rate of the electric field, viz. 200 million times per second) to establish conditions and fill the cavity with power. Once established, the direction of filling becomes immaterial. However, it is important that there is then no reflection at the end: this would cause the power to return and upset the pattern in the cells. So at the (forward) end of the cavity is a water dump, which looks electrically just like the cells before it, but is a device in which the power is dissipated harmlessly. This means that power must be pumped in at a constant rate to maintain a constant field strength in the cells. But the compensating advantage is that as the beam intensity goes up so the efficiency of conversion into proton energy increases: for the planned intensity of the machine it is already over 15 per cent.

The length of each cell is one quarter of the wavelength of a 200 MHz wave, i.e. one quarter of 1·5 m, and as each cell is ringing one quarter of a wavelength behind or in front of its neighbour, the conditions in cells 1, 5, 9, 13 . . . etc. are the same. Cells 2, 6, 10, 14 . . . etc. are also the same, but one quarter of a wavelength

Europe's Giant Accelerator

Power to drive the cavities is generated in a bank of oscillators mounted in auxiliary building 3 on the surface. Each cavity is fed with an r.f. power of 500 kW, which is several times larger than the power that is produced in a big television transmitting station. The various amplifiers are coupled together in groups of two, four and eight to feed the power down to the tunnel.

Opposite, top

The couplings between the various amplifiers necessary to carry this amount of power are correspondingly massive, and the interconnecting feeders are more reminiscent, from the outside, of a water system than an electronic circuit.

Opposite, bottom

Inside, however, the feeders resemble the coaxial cables that connect the aerial to a television set, and the general proportions are similar, as the SPS accelerating system operates at a frequency similar to that used for television transmission.

9/*Acceleration*

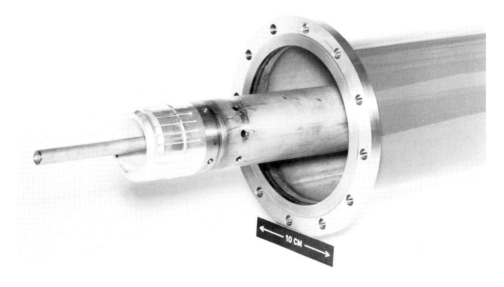

10 CM

different from the first group. This pattern determines the spacing between the proton bunches, the centres of which must be 1·5 m apart, so that the number of bunches in the ring is 4620. Because the speed of the travelling wave is the same as the speed of the bunches, the protons as they cross each cell experience the same accelerating field each time. Over the main part of the cycle, on average they are subject to an accelerating field of 22 000 volts in each cell, and in the two cavities gain about 2·5 MeV on each pass. After 160 000 turns the energy has been raised to 400 GeV.

For this to work, when the bunches have made a circuit of the machine they must arrive at the first cell again at exactly the right moment. The frequency of the accelerating field has then to be tuned precisely to the orbiting speed. However, the fixed dimensions of the structure introduce delays of a fixed time between adjacent cells, so at only one frequency does the difference between adjacent cells correspond exactly to one quarter of a wavelength, and the slippage away from this situation is determined by the difference between the frequency and the ideal, and by the length of the cavity as it is cumulative from cell to cell. Increasing the number of cavities and decreasing their length would diminish the effect but put up the cost, and it was decided to optimize the design on the frequency at transition energy (which we come to in a moment), and adopt a cavity length which meant that in the extreme circumstances at injection (most of the speed change occurs at the beginning of acceleration) with the timing set to match the conditions obtaining in the middle cells, the end cells would be allowed to have a net zero accelerating effect on the protons. This means that the maximum acceleration rate possible is lowest at the beginning, but as we shall see acceleration must begin slowly to give the protons time to sort themselves out into bunches. Accepting this limitation meant that only two separate cavities with their power sources would be needed.

So far we have been considering the protons as if they all revolved at the same speed and arrived together at the first cell at the same moment. They do not. To begin with, the protons are spread out into a continuous ribbon and their energies, hence speeds, vary. If there were no inherent compensating forces at work, it would be impossible to build a machine which could accelerate more than a few particles at a time. Fortunately there are. At the start of the cycle, just a modest power is fed into the cavities, and those protons travelling at a particular speed and arriving at the first cell at a particular time relative to the field oscillation feel little net effect. Those up ahead arrive when the field is against them, so are decelerated; those behind arrive when the field has changed in their favour, so are accelerated. Both groups start to approach their less affected colleagues and bunches begin to form, spaced at intervals of 1·5 m. After a time, those which are ahead have retreated into the middle of a bunch but are now travelling too slowly so will start to lag behind, at which point they will receive an increasing urge forward and will be encouraged once more to approach the centre. Meanwhile, the original laggards will have passed to the front where they will be slowed down. Some protons with the wrong speed/position combination will not be trapped, but most will be gathered into bunches of decreasing size, uniformly spaced out round the ring. After about three tenths of a second, when the bunches are sufficiently compact, the timing can be changed so that the degree of acceleration felt by the synchronous protons (those in the middle of the bunch moving also at the mean speed) can be increased to a maximum of about 70 per cent of the peak field (leaving always some in reserve for the slower members to enable them to catch up), and full power can be fed in to bring the peak field in each cell up to 30 kilovolts.

Throughout the cycle there is a continual exchange between the more and less energetic protons, an individual proton taking about 20 turns to return to the same position in the bunch, only to move away again over the next turn round. There is a certain parallel with what happens in long distance track racing with a staggered start. The bunching gets tighter over the first 10 GeV of acceleration (the length coming down to about 25 cm) followed by some stringing out as the race proceeds, but with a continual exchange between front and back runners.

The protons turn about the centre on a radius determined by their energy and the combination of their speed and the strength of the magnetic field in the ring magnets. If their speed was small in relation to the speed of light and there were only bending magnets in the ring, inside which the magnetic field was constant across the width, the orbit time would be the same for protons of all energies within a range limited only by the width of the vacuum tube, the difference in speed being exactly compensated by the difference in orbit circumference. However, the focusing magnets, in squeezing them together, prevent the protons from following their natural orbit by channelling them into a narrow lane. At modest energies, the more energetic make the turn faster than their less energetic fellows, and as on the race-track lead the bunch. This allows a distinction to be made between protons of different energy, and when a bunch arrives at the cavities it experiences a field which gets bigger as the bunch goes through; the slow ones are in consequence forced to catch up. At high energies, however, those with more energy travel only marginally faster than their companions, but their extra mass causes them to swing out wider round the curves and take a longer route round. As a result, they take longer to make the circuit than the others. Again, this poses no problem, provided the bunch now sees an accelerating field which *diminishes* as it goes through, so

Associated with the power supplies to the cavities is a great deal of electronic circuitry, which adjusts the phase and frequency of the oscillator to the rotational frequency of the bunches of protons round the ring and, in addition, the "height" or amplitude of the accelerating wave. At injection, the average acceleration imposed is very small until the protons have grouped into bunches. Then the power can be raised, but the bunches must see a rising voltage in each cell, as the more energetic protons are orbiting faster than the less energetic. Above 24 GeV, how-

9/*Acceleration*

ever, the reverse is true, the more energetic protons taking longer to orbit the ring than the less energetic, and, in consequence, at the transition energy between the two conditions, the phase must be suddenly changed, so that the back markers in the bunches experience a lower field than the front runners. A subsidiary control room looks after all these problems, but now that the machine is commissioned, there is no need for an operator to be in this room to keep the accelerator running. It is needed only for development work.

the back markers receive a smaller acceleration and the distance they have to run comes down again. The snag comes in between, when they all make the circuit in the same time, and there is no way of picking out one from the other.

This is the transition energy. Crossing transition was perhaps the main bogey of the early builders of synchrotrons, as it was feared that by the time it was safely crossed, and the timing of the accelerating field readjusted, so that bunches met a decreasing instead of an increasing field, they would have spread out so far that most of the protons would be lost. Firmly established in the mythology of the commissioning of the PS is the story of Schnell who, in a flash of inspiration, threw together a circuit in a coffee tin (those were the days of thermionic valves and endless problems of shielding one circuit from another), plugged it in, and transformed the night of pessimistic gloom into a moment of history as transition was crossed for the first time ever. Schnell and the coffee tin are real enough, but there had been a good deal of hard thinking and calculating before the bits that did the trick were assembled. Since then, the theory of how beams behave has advanced enormously, as well as the control of electronic systems.

Nevertheless, even though a great deal of progress has been made since 1959, transition has remained a problem that always has to be treated seriously. As we have seen, the accelerating cavities were optimized for a transition energy of 24 GeV, and the phase control, which sets the middle of the bunch first on a rising, then on a diminishing, field slope, was contrived to make the switch in a fraction of a revolution. Moreover, the moment when this should be done in relation to the focusing magnets' field strength was calculated very accurately. So much so, that during commissioning, although Zettler was present in his control cubicle ready to make adjustments, when the time came to cross transition, he was deep in conversation at the critical moment and did not even notice the event! The beam went through without any modification of the pre-set controls. Phase setting is, of course, not just a matter of setting a few knobs and expecting the system to keep in perfect tune over the 700 million oscillations of the cavities during the acceleration cycle. Probes inside the ring look at the circulating bunches and signal their passing and position across the vacuum tube to the power sources which adjust the frequency and phase to correspond. To aid this operation in the early stages, before bunching has started, a modulation at 200 MHz can be imposed on the beam before it leaves the PS. Such a pre-modulation has not, however, proved to be useful, and bunching can be started perfectly well in the SPS itself.

Once the complex calculations which converted the electronic requirements into physical dimensions were complete and their basic correctness proved in a test cell, manufacture of the shell sections and drift tubes could begin. Ideally, the internal diameter of the cavities should be true to within tenths of a millimetre, but the cost of meeting this tolerance would have been prohibitive. Instead, it was decided to accept the normal industrial tolerances for a rolled and welded vessel of 2 mm, and compensate the errors introduced by adjustment of the diameter and length of the stems which stand the drift tubes off from their pedestals, that in turn are fixed to the walls of the vessel. As each section arrived at CERN, a standard set of eleven drift tubes was mounted along the axis of the vessel and the electrical characteristics were measured which allow the dimensions of the stems needed to bring the characteristics back to the proper figure to be calculated. A new set was then assembled with re-machined stems brazed to the pedestals to a precision of $\frac{1}{100}$ mm. Should subsequent measurement show a slight deviation from the ideal, a correction could still be made in the next section. In this way, the overall behaviour of the cavities could be brought very close to the ideal.

In parallel with this work came the design of the power supplies. A big television transmitter has an oscillator power of 10–20 kW, but each cavity had to have a radio-frequency power of 500 kW. Industry was asked for its proposals. A number of designs were submitted based on 2, 4, 8, and 16 thermionic valves. The greater the number of valves, the more nearly they approached the size of valves already developed, but also the more complex became the coupling linking them all together. Finally, a German proposal was accepted incorporating eight valves in the final stage of amplification. The really tricky part of the design of these valves is to stop them resonating at a frequency different from the oscillator frequency, at which point instead of the energy pouring out into the external line, hot spots develop and the valve burns out. There is no exact science available to guide the designers in their efforts to suppress these unwanted modes. It was only after a great deal of head scratching, and trial and error involving quite a few burnt valves, that a design was evolved which met the specifications.

Because there is no timing to be done on the cavities themselves, both the power sources and their control system could be located on the surface well away from the machine. The main amplifiers are located on the main floor of auxiliary building 3, while in a special shielded cage is generated in a quartz oscillator the basic frequency tuned before amplification in frequency and phase to correspond to the accelerator sequence. At the beginning of the cycle the two cavities oppose each other, so that zero acceleration takes place while the protons group into bunches.

The power is fed down to the cavities in two pipes, blown-up versions of the coaxial cables which link a television aerial to the set. For the SPS, the outer conductor is of aluminium with an inside diameter of 34·5 cm, and the inner conductor is of copper 15 cm in diameter. Sliding joints take care of the different expansions of the two materials as the temperature changes. At the ends, ingenious couplers match the transmission line to the cavities and minimize the losses. One of the more awkward engineering problems was to isolate the vacuum in the cavity from the air-filled transmission line. Beryllium oxide was chosen for the cylindrical windows, as it has a lower dielectric constant and 10 times higher heat conductivity than the more usual aluminium oxide. But this did not provide the complete answer. Inside the cavity, locally released electrons bombard the surfaces of the structure under the influence of the r.f. fields, an effect which is suppressed at atmospheric pressures because of collisions between the ions and the air molecules. The effect in the cavity was to boil off the gold from the contacts near the entry, which then deposited on the window, and finally created a short circuit. A partial solution has been to coat the region surrounding the windows with titanium, which is less prone to give off the offending electrons; even so, provision is made for rapid changing of the windows by the use of quick release vacuum joints so that outage time can be cut to a minimum if the effect persists.

At a time when colour television and electronic watches are a commonplace, it is easy to underestimate the complexities of an installation such as the SPS accelerating system. Yet the very simplicity of the accelerating cavities is a triumph of theoretical analysis and calculation, while the combination of fine control in phase and frequency, married to outputs of such high power, demonstrates an impressive mastery of r.f. engineering.

10/Machine Management

Ten kilometres of machine components, 170 000 kilowatts of power, maybe 20 000 situations to survey, 6000 values to adjust, and no means of reaching half the equipment without going through customs checks or traffic lights. This was the daunting prospect facing the Control System Working Group when it got down to devising a system for managing the largest machine in the world, before it was even designed.

Yet the huge complexity of the task was not the dominating preoccupation in the early days, but the need to guide the beam in the main ring with such precision that the margin allowed for errors could be cut to a minimum. As we have already seen, the protons in the beam are not lined up one behind the other, but extend over a certain area; nor do they move as a constant parallel stream, but balloon out and in, under the influence of the focusing magnets. The envelope which contains them is termed 'the closed orbit', and this, together with the spread in width caused by their spread in energy, defines the minimum theoretical aperture through which the beam can pass. Any deviation in the closed orbit resulting from imperfections in the uniformity of the fields in the bending or focusing magnets, or in their alignment or tilt, or from stray magnetic fields, adds to the size of aperture needed. Past practice, when determining the cross-section of the main ring vacuum tube, had been to make full allowance for such errors and add a margin of safety. The additional cost of doing this in the SPS, where the estimated accumulated errors were 32 mm in the horizontal plane and 19 mm in the vertical, would have been considerable. The decision taken was to devise a control system which would be able to compensate for almost all the errors expected, leaving margins of only 10 mm in the horizontal, and 5 mm in the vertical, planes.

A catch-phrase of the time was 'the cybernetic machine', where control of the closed orbit would be made automatically. Instruments would measure the deviations from the closed orbit on one cycle, and the control system would make the necessary calculations and feed information to the correcting magnets to bring about an improvement on the next cycle. In the event, the precision of both the field and alignment of the magnets has proved to be so good that automatic correction of the orbit could have been dispensed with. It is all the other virtues of the control system that have been shown to be crucial to the commissioning and operation of the accelerator.

A second system of auto-control was concerned with beam stability. To a large extent, the corrections for the distortion of the closed orbit stemming from the causes mentioned above can be calculated in advance, but it is much more difficult to anticipate instabilities of the beam which might appear at high intensities, through an interaction between the beam and its environment. One source of these can be the currents induced in the wall of the vacuum tube by the passage of the proton bunches. Such currents depend upon the position of the beam relative to the wall, and they flow at a speed that is less than the bunch speed. In a particular combination of circumstances, the currents induced by one bunch can influence a bunch behind and move this bunch closer to the wall, amplifying the effect for the ones which follow. The distortion in the orbit can then build up progressively. There can even be resonance effects within a single bunch, because of the continual exchange of positions of the front and back markers. Front markers, as they move back, can find themselves responding to currents which they have produced, and once more the effect can be cumulative. Over several turns round the ring, such resonances can be disastrous, creating serious beam loss or even its total destruction.

What would be ideal would be a system which, having detected the start of an instability, applied a damper a few metres downstream; but when dealing with a beam of particles travelling at the speed of light, which is the maximum speed at which any signal can be sent, such a procedure is out of the question. Instead, after an instability has been detected, a signal is transmitted straight across the machine, and a correcting electric field is applied when the offending protons arrive, having taken the longer route round the circumference. In practice, the accelerator has turned out to be a very 'clean' machine—as someone remarked, "it had read the text-book"—and although some instabilities must be expected at very

Opposite
The first preoccupation of the Controls Group was to devise a system which would automatically control the position of the circulating beam in the vacuum tube, as the cross-section of this tube allowed very little margin for error. In practice, the machine was so steady that automatic control of beam position and beam stability proved to be a refinement rather than a necessity. The photograph shows one of the monitors that could be moved into, and out of, the beam path.

high intensities, they were little in evidence to begin with. When they appear the control system will be ready for them.

It needed little study to conclude that the sheer size of the control task would mean that computers would be heavily involved. When the PS was built, computer technology was still in its infancy, and there was no alternative but to bring cables back from every piece of equipment to a control area, where rows upon rows of electronic racks fitted with knobs and dials allowed a team of operators to rush around noting figures, correlating information, and making adjustments. Separate control rooms were needed for the proton source, the linac, the power supplies, and the r.f. which were, in turn, linked by cable and by telephone to a central control room. During commissioning, it was often easier to go and talk to the people in one of the sub-control rooms when things got tricky, than to try to sort out the situation on the telephone. Quite simple, but vital measurements, could be made only by manually adjusting certain values, assessing the effects by reading innumerable dials, then making lengthy calculations. Computers were added later to handle much of the routine data and fault signals, and to set up interconnected functions when a new operational sequence was required, but they were there as aids to the running of an essentially manually operated machine. With a much larger machine such as the SPS, the cost of cabling alone would have been prohibitive if the same approach had been adopted, quite apart from the huge management problem of how to arrange and manipulate, with anything like the accuracy required, all the control items necessary.

The problems may be regarded as not dissimilar from those of managing an industrial enterprise, and many a business limping along with an out-dated management system might reflect on the principles and techniques employed in the control of the SPS. The machine has a very large number of separate elements, which individually need to be supplied, regulated, and observed. Many of them must respond to what happens elsewhere, their actions must be co-ordinated; some must be able to adjust their mode of operation according to a central policy decision, which in turn is conditioned by the demand; others have routine jobs to do regardless of what is going on in other parts of the machine; incorrect functioning can cause troubles ranging from a mild irritation to a complete stoppage of the whole complex. At the top, someone must decide what the policy is. This policy must take into account what the customer (the experimental physicist) wants, and what the machine can supply. The director must be able to know at all times in what state his business is, but neither be so limited in his information that he is unable to make reasoned decisions, nor so overburdened with detail that a serious block develops, in which reactions are slow and no-one can show initiative. In

10/Machine Management

addition, the whole management system must be sufficiently flexible to accommodate changes in the market requirements, developments in techniques, and the unexpected effects of external influences or internal interactions.

In view of the intricacy of the whole system, it might have seemed logical to put its design into the hands of a professional administrator, who in technical terms would be a control engineer, or as computers were to be an integral part of the management system, a computer expert. (It would evidently have been quixotic to put in charge, say, a power supplies specialist or a purchasing officer, the technical equivalents perhaps of a financier or a lawyer.) Instead, the man chosen as convenor of the Working Group, who then became head of the Controls Group and eventually head of the SPS Division when the machine became operational, was an accelerator expert, Michael Crowling-Milling. From building radar for ships with Metropolitan Vickers, which gave him a taste for sailing small boats off the north coast of Wales, as well as having everything trim and ship-shape, he moved over to accelerators, designing one of the first 10 MeV electron accelerators used as a source of X-rays in hospitals, and then assisting in the design of the proton linear accelerator of the CERN PS. Back to electrons with the design of the injector for the DESY accelerator in Hamburg (which was handed over two months ahead of schedule), he then joined the NINA team at Daresbury, Cheshire. There he designed the injector, and progressively took charge of beam instrumentation, power supplies, vacuum, r.f. and commissioning. Following the successful operation of NINA he turned his attention to machine physics and new developments. On the side, he was collecting and rebuilding veteran cars. Reserved and courteous in manner, his slight stutter may have helped to make him a good listener, and to get his thoughts clear before expounding. He not only knew accelerators from the inside, but also knew the operators' problems, and knew the customers, the physicists.

Computers were by now being used extensively for the control of accelerators, and the dangers not only of the machine, but also of the personnel being run by an automaton were beginning to show. It was agreed that the SPS was not to be run by a rigid, inflexible system that required the complete behaviour of the machine to be known and understood before the computers could be programmed, and in which any change would need a new program to be written, read into the machine, found to be faulty, debugged, and compiled once more, while everyone stood around and waited. The system had to be capable of evolving with the design of the accelerator up to and beyond commissioning, it had to give the operators the information they wanted when they wanted it, and it had to carry out a wide variety of tasks on its own without operator intervention, except in case of need.

Europe's Giant Accelerator

Michael Crowley-Milling, an accelerator man from the Daresbury Laboratory in Cheshire, was made head of the Controls Group, and later, after integration, head of the SPS Division. He combined a special knowledge of beam control systems with a broad experience of accelerator design and operation.

10/Machine Management

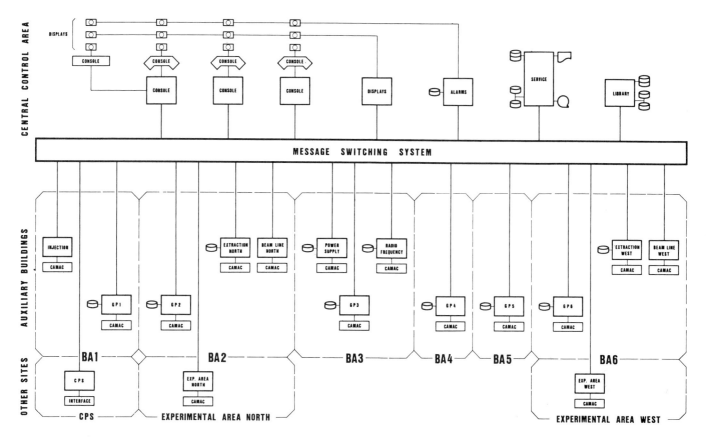

The machine is controlled entirely through computers, some of which look after all the operations in a given area, while others are dedicated to specific tasks. They are all identical and of equal status, and are interconnected through a message transfer system.

The computer experts could provide all this once individual demands for control facilities were formulated, vetted, the superfluous proposals rejected, and priorities set down.

By the time agreement for the SPS project to go ahead had been given, the Working Group had evolved a skeletal scheme for the control system, and the Controls Group could begin to put flesh on the bones. What was envisaged was a central control computer linked to a number of small computers scattered around the site, some of which would be dedicated to specific tasks, such as beam ejection, while

As much of the data and sub-routines are stored local to the computers, the message transfer system is not overloaded by the need to communicate continually with a centralized data-base. Moreover, even if a computer fails (with the exception of the computer controlling the main ring power supplies) the machine will continue to run for a time on the basis of the last information received.

others would be general-purpose computers looking after all the tasks in a given sector. All the changes in operating modes, the surveillance, and the transmission of data from one part of the machine to another, would pass through the computers. For equipment distributed over a wide area where it would be uneconomic to make direct connexions, a multiplexor would be incorporated (this resembles a telephone network through which individual items of equipment are called up from a local exchange when something is to be done).

10/Machine Management

At the Rutherford Laboratory, a small control system using an interpreter had been successfully demonstrated. An interpreter is able to accept a message given in plain language, and translate this into a number code which the computer is able to understand. In its turn, it will accept numbered codes from the computer and translate them into plain language as a message to the operator. The adoption of this technique was to be of great significance not only in simplifying the work of the operators, but also in helping the engineers to develop and test their equipment through the system, using the same language as the operators. From the beginning, the control system was not something that had to be thought about once equipment was built and installed, but was an integral part of the equipment, to be considered from the outset and used as a tool during development and testing.

Once the system looked tidy, a preliminary enquiry was sent to computer manufacturers for their proposals for a complete system—of computers, of the software that tells them what to do, and of the message transfer system that allows them to talk to each other and to the equipment. The replies were disappointing; it became clear that no one company was able to tackle the complete job, and quite a few did not even understand the problem. It seemed that the computer manufacturers were not interested in the message transfer, and message transfer specialists were not able to provide the computers. One point that emerged, from the option given to the companies to propose either one central computer or a group of small computers, was that where the two possibilities had been considered, the group of small computers was the cheaper.

It was decided to divide up the job. A new call for tenders was made for a battery of identical computers, which would meet the specifications laid down by CERN. There was to be no master computer. All computers would have their own interpreter, and all had to be capable of talking to each other through a message transfer system which would be specified separately. The computers would be expected to feed into standard input/output boxes called CAMAC, developed primarily for high energy physics research, which, in turn, would feed adjacent equipment, or the multiplexors talking to the distant equipment. The specification went into considerable detail, and it was not expected that any manufacturer would be able to come up with a ready-made system that satisfied all the requirements.

It was then with some surprise and considerable pleasure that the tender jury studied the offer of a Norwegian company that had made a speciality of ship control by computer. Their offer was based on a development of their Nord–1 computer, which was to be ready soon and would be called Nord–10. If the development had meant a quite new machine, needing a new set of software to make it work, it would never have been considered. Users, and CERN in particular, have learned to be extremely wary of such developments, as it is often years before the software is finally tamed, whilst the hardware, if properly designed, can be made to function reliably quite quickly. The Nord–10, as far as software was concerned, would be completely compatible with the Nord–1, which had been operating for many years in situations where reliability was paramount. If a system breaks down when a ship is in Japan, an engineer has to fly out there and is unlikely to come ashore until the next docking, an expensive business for the firm. The major change in hand was to redesign the hardware to take advantage of modern electronic developments, such as integrated circuits, and produce a model that did not have to withstand the rigours of ship-board treatment. When the Group came to the Finance Committee for the adjudication of the Norwegian contract, the scepticism felt by delegates from some of the bigger countries was ill-concealed, and CERN was warned of the risk it was taking in turning down the offers of the larger manufacturers. For the record, the first Nord–10 would have been delivered to CERN on exactly the date specified, but for a strike of customs officials in one of the big countries en route. It, and the 24 others in the main system, have in operation given very little trouble.

Specifying the message transfer system was not an easy task. How many messages would have to pass, how long would they be, what priority should they carry? These were questions which could only be answered precisely when the accelerator and beam lines were fully designed. It had been possible to decide on the number of computers on the basis of the number of sectors, specialized facilities, and control functions, and a reasonable estimate could be made of the size of memory each would require. Moreover, additional memories could be added if the need arose. The message transfer system had to be designed so that it would be able to cope with any peak traffic. Estimates from the various SPS groups were carefully analysed, but in the end the quantity had to be a matter of intelligent guesswork, conditioned by the decision to provide a library with mass storage facilities. As to message length, it was decided to fix a maximum package size, and to give each package a title which contained information on the source, destination, priority, etc. Long messages are sent as a series of packages, with only a small time penalty.

Central to the transfer system is the message transfer computer, also a Nord–10, which takes in messages coming from all the other computers, and re-distributes them to their destinations. In the process, it makes checks on their authenticity, but they are passed on without modification, and priorities are respected. It fulfils the same functions as the internal post room in a company, the other computers

10/Machine Management

taking the place of the offices of the various departmental heads, who need to communicate with each other as well as with the management. First priority is given to the essential mechanics of the communication system, checking that offices are open, for example, and that a message can be received. Next come the commands that require immediate execution; after that the short messages; then the packages that take up more time. In the other computers, a scale of priorities has again been laid down, but there is no sitting back waiting for actions initiated to be carried out. As soon as there is a pause while, for instance, data are collected, the computer goes on to the next tasks in order of priority. Top priority is again awarded to the system; it is no good expecting messages if you have left a "gone to lunch" sign on the door. Incoming messages can be put into the in-tray (the buffer) without delay. These are not allowed to pile up in any haphazard manner. Demands for immediate execution are dealt with at once, and those of less priority afterwards. When outside affairs have been put in hand, the computer can see to its own departmental business, in particular making the rounds of the equipment to see that all is going well. Questions from the central control desk are next on the list. Unlike in business, a demand for discussions with top management can be agreed to only when it does not interfere with departmental work. At the bottom, comes local interactive discussion: the chap who drops in for a chat has to wait until there is nothing else more pressing to be done. In the last category, when there are likely to be a number of callers of equal importance—for example, in the experimental areas where a number of people may be wishing to interrogate the system simultaneously—time sharing is introduced. Because the computers are able to interleave their tasks, waiting times are minimal, and internal checks ensure that a command to perform a repetitive job, such as reading out a particular value, is not absent-mindedly left running. After a pre-set time the action will stop, unless there is a renewed instruction to carry on.

Not all messages need to travel at the same speed, and it was specified that computers would be linked by cables originally developed for television relay work. These consist of a 'video' pair and an 'audio' pair twisted together. The expensive (video) pair can send signals at the rate of several hundred thousand bits per second; the cheaper pair at tens of thousands of bits per second. (A bit is a yes/no signal; combinations of bits make a word, which in turn represents numbers or letters that form the communication language between computers.) The video pair is reserved for the message proper, while the audio pair takes care of recognition and alarm signals. The specification finally identified the maximum mean rate as 30 000 words/second and the maximum package length as 64 words, plus a four-word header. In addition, it had to be possible to transmit a short

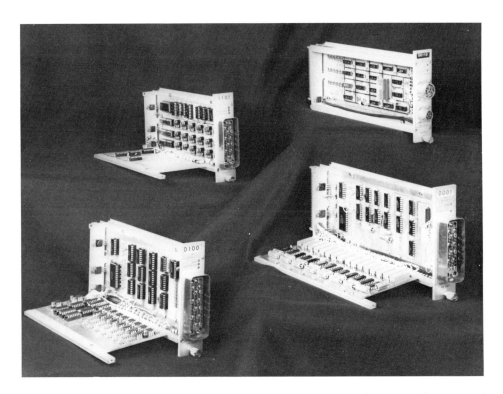

All the adjustable controls and the monitors which indicate the status of a piece of equipment are coupled through electronic units to one or other of the computers. Standardization of these units in both overall design and plug connections produced an economic and homogeneous system.

message from one computer to another in a total time of 5 milliseconds. The possibility of having two identical systems to guard against failure was included. The computers would transmit alternately on each system, but if a computer found it was getting consistent errors on one (the words always contain check codes), it would route all its messages through the other, and signal the malfunctioning to the control room. Eventually, only one was installed, but a spare computer is always on stand-by.

Three companies made offers for the supply of the message transfer system, and the contract was awarded to a French company, which resulted in some delightful multi-lingual jargon evolving as the computer equipment and most of the system was described in English. In an information bulletin issued to users we find,

10/Machine Management

'control over the various functions of the voies of the coupleurs is achieved by signals on the bus groupe. These signals are generated by the logique de groupe or interface satellite circuits of the coupleurs.'

Most computer jargon is unintelligible to the uninitiated, so such changes of tongue could no doubt be taken by the engineers in their stride. In spite of the difficulties of defining the message transfer system, the novelty of the job, and the early stage at which it was ordered, the equipment arrived on schedule, and was in service in about one-third of the time that had been allowed on the programme. Moreover, it has proved well up to its task, and there is even some spare capacity in hand, though this will no doubt get used up.

From the time taken to pass even rapid messages, it is clear that the control system is not directly involved in such high-speed operations as the synchronizing and phase shifting of the r.f. (Top management does not get involved in departmental affairs when things are going right.) The control system is there to make changes when needed, and to keep a check that all is well. Only one computer —the one which controls the magnet power supplies—going down will cause an immediate stoppage of the accelerator. Any of the others can go down, and the accelerator will continue to function for a time on the basis of the last instructions received. Should one of them start sending irresponsible messages, internal system checks will detect the derangement, and ignore the instructions.

Most of the key operations, such as the programming of the r.f. accelerating pattern, are pre-set to follow a certain timing sequence derived from the cycle shape that the operator has demanded by specifying the desired rise and fall of the field in the main ring magnets. A standard timing signal at millisecond intervals with additional intermediate codes is piped round the ring through the multiplexor system. This is triggered by a series of signals from the PS announcing that it is ready to inject, the last one being sent to the SPS, via the injector pulser, 50 microseconds before injection starts. The r.f. will then come in at the time that has been pre-selected, while the locking of the frequency on to the speed of the circulating bunches, and the adjustment of phase, are controlled by a very high speed loop linking the power source, beam monitor, and cavities directly together. What has to be done as the cycle proceeds is to read out from the memory banks, with which the computer has previously been loaded, according to the operator's instructions.

The exception to this general policy, as we have said, is the control of the magnet power supplies, as the important characteristic is not the voltage which is applied, but the magnetic field in the bending magnets, and in the two sets of quadrupoles. As the temperature of the magnets changes, so the field varies for a given applied

Europe's Giant Accelerator

voltage, and compensation must be made. In series with each of the three groups of magnets in the main ring—the bending magnets, the vertical focusing quadrupoles, and the horizontal focusing quadrupoles—are test magnets of similar design but with measuring coils inside. In the computer dedicated to the magnet power supplies, there is a table which is read every 10 milliseconds, and a new set of figures sent off by direct links to the rectifier stations indicates the rate of change of voltage desired over the next step. Every 60 milliseconds, signals are sent to the computer showing the actual magnetic fields measured, and it then calculates how it should modify the table for the next cycle to arrive at a better fit. From a new start, about 10 cycles of the accelerator are needed to get the fit absolutely right, following which any slow drifts will be corrected more or less as they arise. The rise and fall of the applied voltage follows a complicated pattern in each of the three groups, and the calculations necessary to maintain the demanded pattern keep the power supply computer pretty busy. Should it fail, then the accelerator stops, and it is necessary to replace the computer by wheeling another into its place, and changing over the connections. A copy of the memory held in the main library file is then read into the replacement computer, and the accelerator can start up again.

Automatic control of the closed beam orbit is a faster iterative process that can be applied from one cycle to the next. The beam is at its most delicate on injection, when remanent field effects in the magnets can distort the beam path. Small correction magnets are installed before each of the quadrupoles, which can be separately powered to deflect the beam, either vertically or horizontally. Near them are beam position monitors which signal to the control system the position the beam takes up. This information can either be fed to the automatic control, or displayed to the operators. If the machine is on automatic, a computer will calculate what changes need to be made to the correction magnets before and after each monitor, to straighten the beam, and bring it into the centre line. On the next cycle the beam will run down the middle all the way round. Alternatively, the operators can set the individual power supplies manually from a control desk in the central control room, through the computer system. Similarly, the Q-values can be measured in steps by following a series of instructions given by an operator who, on reading out the oscillations provoked by the changes he has made, can use the computer as a calculator, or he can ask the control system to do the whole measurement on its own.

While the computers and message transfer system were being built, the Controls Group was engaged on several major jobs: devising the language to be used in the interpreter, developing the displays and control elements for the operators, informing the various SPS groups on how the system would work, helping them to write the programs that would allow the computers to look after their equipment, writing programs for the internal communications within the system, and establishing standards for the interface equipment of the multiplexors to minimize the number of different control boxes that would be required for the multifarious duties around the machine.

No language was available which could be readily adapted to the specific requirements of the SPS interpreters, where the main emphasis is on control functions rather than mathematical computation. It was decided, therefore, to write a new one which would be as easy to learn as possible, and would simplify the problem of writing programs for jobs to be executed in a number of computers. It was given the name Nodal, to acknowledge its partial derivation from DEC's language Focal, while forming the acronym NOrD Accelerator Language.

As every computer was provided with its own interpreter, as soon as a local computer was installed it could be used for checking out the equipment to which it was assigned. The computer is coupled to one or more standard CAMAC units, to which nearby equipment can be directly connected, whereas distant equipment is coupled through a multiplexor. A single CAMAC crate can control four multiplexors, each of which can serve 32 stations. The CAMAC crates contain a number of special-purpose modules, whose job is to verify that the action required is reasonable and meaningful, and to translate the computer output into explicit actions, such as counting the number of pulses received, or setting a voltage to change at a specified rate. Also, it vets the information it receives, and translates this information into terms the computer understands. If the equipment concerned is connected via the multiplexor, the control unit in the multiplexor will transmit the appropriate code and function round the stations, which can take up to eleven plug-in modules each. A decoder will see that the appropriate operation is carried out on the appropriate equipment, whether this be for example, setting a switch or reading off a value. All the other modules and stations will ignore the message. A series of simple modules, equipped with standard plugs and sockets, were designed able to satisfy the majority of the requirements of the engineers. For the rest, special-purpose modules were built. In addition, a number of special interface devices were designed which, for example, converted a voltage reading into a number code and vice-versa.

The engineers were at first presented with a typewriter, and later with a display, connected to a 'black box' (the computer), and, provided they had learned the basic command language, and knew the codes of the crates and sub-units connected to the controls in which they were interested, they could test their

Each computer is headed by an interpreter which allows direct communication in a basic English language developed for the system. As a result, all the computers could be used separately for proving out individual control circuits and the routines written to undertake a pre-determined control sequence.

equipment step by step. Once they were confident that each bit of equipment was functioning correctly, they could move on to the next phase, to check out the programs they had prepared, which combined a number of instructions into a logical control sequence. Some of these are relatively straightforward when expressed in words; for example, "Read the position of the beam in the beam monitor no. X." However, such an instruction must be broken down into a series of command actions to send the right message through the multiplexor to the appropriate module in the relevant station, define the time at which the measurement is to be made, collect the information, and then feed the result into the system which is able to present it in a comprehensible form. Writing such programs is quite a lengthy business; getting them right is even lengthier. Some of the programs are extremely complicated; the complete program concerned with setting the ring

10/Machine Management

Behind the main control room are the computers which look after the message transfer system and those dedicated to the task of translating information into various types of display, so that the operators can 'see' the performance of the machine and its components, and study data already collated in a readily assimilable form.

magnet power supplies looks like a small book on its own when typed out. The time and effort necessary to get it perfect was considerable.

Although the Controls Group was there to help in writing these programs, it was the engineers who had designed the equipment who were responsible for doing the job, and checking them out. It was at this stage that the wisdom of fitting each computer with its own interpreter became evident. All over the site, engineers could be found typing away, observing what happened, correcting their programs, rushing off to check equipment, and returning to their typewriters to try something new. This would be going on at all times of the day and night, all the Groups getting on with their own affairs without interfering with each other. Many of the operations have, of course, functions which depend upon the state of the machine, and a fundamental policy decision was that nothing about the state was secret. Anyone could interrogate any parameter once the appropriate controls had been hooked in, so there was never any excuse about not knowing what was happening. How different this policy is from that adopted in most business management operations! Not everybody, however, could make alterations to a particular setting, or the logic of a program. Codes identified the people responsible for these, and only by presenting to the computer the code and his personal pass card could an individual make modifications once a piece of plant had been commissioned and its *modus operandi* established.

So far we have not mentioned the second stage in the link between the computers and the CAMAC crates, a data module; which, in the jargon of the system, formed part of a distributed data-base. In other systems designed for similar control operations, all the information relating to the machine, including the programs for carrying out the detailed actions at any point, was held in a huge central library. With such a system, the queuing that goes on to get at the data creates bottle-necks, and when any modification has to be made a new program has to be fed into the system, with a high probability that errors will be introduced at the same time. With the distributed system, programs could be developed and debugged at the terminal connected to the relevant equipment, and then loaded into the computer's memory. (A copy of this program could then be filed in the central library in case it got lost.) The handler which headed the data module would be informed of its existence and its coding, and a simple command from then on would gain access to it. There was no unnecessary blocking of the main communication lines, and these could be reserved for transmitting data needing to be collected from other parts of the machine and other messages, while the central library could concentrate on filing and distributing data of general significance.

From that point on, all that was necessary to carry out a given operation was to type in at any of the terminals a 'plain' English instruction in the Nodal language, and the appropriate sequence of actions implicit in this command would follow. This language, common to all the interpreters and understood by dozens of people working on the SPS, was designed to be as clear as possible. Here is an example:

$$\text{SET Mag } (25, \neq \text{Swi}) = 1$$

This is understood by the computer as meaning: "I must send out on the multiplexor an instruction to carry out an action (SET) on magnet (Mag) number 25, relating to the property ($\neq$) switch (Swi), viz. turn on ($= 1$)" If the instruction had been to switch off, the code would have been $= 0$. A list of such commands, written as a series of numbered lines, can be referred back to if needed and repeated, and an engineer can build up an operational sequence that can be called up at will.

Such step-by-step procedures were very necessary to commission the various items of plant equipment, and the typewriter is an invaluable device for transmitting messages and receiving written answers, but it falls far short of the needs of an operator in a central control room who wants to 'see' what is going on, look at the shape of performance curves, or set or interrogate equipment, without first having to look up in a directory what the appropriate description is. A very considerable effort therefore, was made to devise a control console, which would allow one person to look after the whole machine without having to live in a sea of paper. Many a manager will sympathize with this ambition.

One of the first of the display devices to be developed was the presentation of graphical and tabular information on a television screen. By the use of codes which can cope with lines, characters, upper and lower case, figures and a certain number of symbols, graphs, tables and mimic diagrams can be presented. The use of different colours allows a further coding to be introduced, so that, for example, on a table of parameters, the current ones used in the machine can be shown in one colour, whilst a new set of values the operators are setting up will appear alongside in a different colour. For graphical information, one set of data can be presented in one colour, another set relating to the same time interval in another. Black and white screens, which take less computer effort, display data on the main machine parameters, and this can be displayed simultaneously at a large number of output stations. Equally, they can be used to display other information, if required. Televiews of entry doors, and of luminescent screens which mark the passage of the beam, can also be sent back through direct video links to dedicated black and white screens in the control room. Graticules of different grill dimensions can be added in colour on the colour screens, and cursors set by the movement of a rolling ball on the control desk. The cursors can be used to identify a portion of a graphical display, which the operator wishes to have blown up so that he can study the shape of the curve in detail.

The rolling ball, however, can do more than just set cursors. It can be used to control the position of a spot, which is used to identify a particular piece of equipment within a system displayed as a mimic diagram on the screen. Moreover, the value of interest relating to this equipment can be displayed in a little square alongside it, and if this value has to be changed, the spot identifies where, and the square shows the reading.

Adjustment of any parameter can be made through a single knob, which can not

Three independent control consoles in the main control room allow special studies to go on, in parallel with machine operation. Each is provided with coloured and black and white television screens, a typewriter, rolling ball, knob and touch panel through which individual components or control programs can be called up.

only regulate a vast number of settings, but also can be asked to do this in a variety of ways. A cunning set of gears and clutches underneath allows the knob to behave as a smooth regulator, as an indexing control with a series of set positions, or as a spring-loaded knob, so that it returns to zero when released. It can even have inbuilt 'feel', so the operator has to make an increasing effort to go on turning.

During the development of these devices, when the only things that could be displayed were simulated situations, it was undoubtedly the most popular toy in the SPS, so pretty to play with and fascinating to watch that it was difficult to believe at times that it was for a serious purpose. But as soon as commissioning began, the ease of control and the power of the display system left no-one in doubt that the game had been worth the playtime.

One other facility designed to simplify the operator's task has also proved to be of the greatest utility. Instead of the various sub-programs associated with the manipulation of the different items of equipment being stored in some arbitrary order, they are arranged in the form of a tree, with boughs, branches, twigs, and leaves. A tree, moreover, that is a living thing, which can grow new branches and discard old ones. The whole of the machine is broken down into a series of main functions, and these in turn are subdivided into classes, and so on down to particular groups of equipment. The operations that need to be carried out can also be classified, this time into a relatively small number of general categories—set, read, display, and so on. Mounted on the operator's desk is a small rectangular translucent panel, on which are displayed up to sixteen simple descriptions like the section headings in a library index. Touching the appropriate section can cause a new display to appear, such as the subjects covered in that section; another touch the book titles on the chosen subject appear; another touch the chapter headings in a particular book are seen; touch again and the paragraph headings are presented. Just a few touches on the SPS screen and the particular equipment of interest can be brought to the fore. Using the same screen, the type of action to be taken can also be selected, and the operator can turn his knob, roll his ball, send commands through his typewriter, or, if the action has already been previously defined and a program written, just sit back and let the computer get on with it.

Touch screens have been on the market for a number of years, but each has its snags. Consequently, it was decided to try something new, and to activate the 'button' by the change in capacitance caused by a finger approaching the screen. Such devices have been successfully developed for single on–off switches, and are becoming popular as domestic light switches and lift controls. The requirement here, however, was for a battery of transparent buttons, behind which characters could be written on a cathode ray tube, fed by the display computer. After some anguish on the part of the manufacturer chosen to reproduce the complicated pattern of invisible conductors printed on glass, a successful process was developed. As a result, the operator has now in front of him a solid glass plate on which he can jab his finger without fear of doing irreparable damage. One further elaboration was added, to complete the illusion of a button; a loud clonk is generated when the operator makes contact, to give him the satisfaction of knowing that the computer has noticed.

Access to areas which could be dangerous is only possible by following a routine which ensures that the machine is off, the person wishing to enter is authorized to do so, and if there is a radiation hazard the dose he receives is measured and recorded.

There are three control consoles in the central control room, any one of which can act as the principal control panel. The other two are, however, fully operational, and can be used for equipment interrogation and adjustment at the same time. The system is aware of this, and has built-in safeguards to prevent two people trying to do two different things with the same equipment at the same time. The danger is not so much one of competition or confusion between the operators, or unlawful actions on the part of an unauthorized person (this is taken care of by the pass-code precautions) so much as the danger of mixing a message while data are being taken from a memory and losing the original program. To guard against this, the computer will make sure that one instruction has been fully executed before going on to the next. Also, in the control room is a single station devoted to alarms which signal any malfunctioning of the machine, or the breaking of interlocks which it would expect to be closed.

Personnel protection is taken very seriously, and access to certain areas which may be dangerous because of radiation or high voltages, such as the machine tunnels, is controlled by a combination of safety checks. Those areas considered to be hazardous are sealed, and can only be entered with the permission of the chief operator, and then only under strict surveillance. Before the accelerator is brought into operation after a shut-down, an announcement is made in four languages, and a systematic search is carried out to make sure that all the people who were working there have left. The access gate is then locked, and the area classified as red. Should someone have escaped notice, he can stop the machine operating by leaning on one of the emergency stops. If an engineer is obliged to enter a red area, he must first insert his identity card into the appropriate slot in a key panel fixed near the entrance gate, and call up the control room to explain his reasons. If these are considered to be adequate, the machine is shut down, and his name will appear above one of the keys which he is now able to remove.

The accelerator cannot be started again until this key is replaced. Watched through the closed circuit television system from the control room, he is able to open the gate and enter the area. If it is a radiation area, he must also take a monitor from the charging rack by the gate, insert it into the reader so that the zero can be checked and logged against his name. On returning, he must go through the same operations in the reverse order, when the dose he has received will automatically be displayed and centrally recorded. Should the local computer be out of action, a routine is laid down which requires the operator from his post personally to check each movement. During construction, before the red areas were established, the entry gates were manned, and access to the tunnel was controlled by guards who, in exchange for authorizing passes, handed over a badge which had to be returned on leaving. Only by special arrangement could people go down to the tunnel by one shaft, and come back via another.

For verbal communications, the site is wired for public address, intercom and headphones, and three UHF frequencies have been allocated for short distance radio transmission.

Any management system must be judged on the efficiency with which the business is run, and on the reactions of the staff. The speed of commissioning of the SPS was testimony enough to the efficiency of the control system. You have only to stand for a few minutes at the elbow of an operator to appreciate how smoothly new instructions can be fed in, parameters altered, or the state of the machine surveyed. There is no congestion on the lines, and the routine actions for keeping the accelerator running go on quietly in the background, without intervention from the central desk. As one cynic remarked soon after the beam had made its first tour of the ring, it is so easy to do experiments on the accelerator's behaviour, the tendency is to go on doing them without stopping to digest the results.

For the operators, the control system is a joy; of equal significance is the enthusiasm it has generated in all the engineers who made use of it during the development and installation phases. People need no encouragement to explain to the passing visitor what they personally were responsible for, the routines they wrote, the little bit of the overall system they created. Men and women of all grades watch it in action with satisfaction, identifying their own program when it comes into play, when their own part of the complex is displayed on the screens. There is a feeling of pride as the buttons are tapped, and their own bit of plant, perhaps deep down in a tunnel five kilometres away, responds to the quiet instructions spelled out. All this says a great deal for the system, and the people who devised it. It also says a great deal for the efficiency of communication between the Controls Group and the other Groups during its development, and for the soundness of the policy that it should be accessible to a wide range of people, and not remain the coveted preserve of a select few.

The SPS control system has made a great jump forward in the control of accelerators, but its significance is not confined to high energy machines alone. It has paved the way for a new approach to centralized plant control, and it embodies principles which have a direct relevance to the management of a wide variety of undertakings.

11/Start-up

Well before the SPS was finished, plans were made to cover the transitional stages from building to commissioning and from commissioning to operation. In the spring of 1975 it was announced that three of the Group Leaders would be assuming responsibility for the project from the beginning of the following year: Bas de Raad would be taking over the accelerator, Giorgio Brianti would look after the experimental areas (see Chapter 12), and Lévy-Mandel the civil engineering work. Construction of the north area was only just getting under way and most of the building work over the last two years of the SPS programme would be concerned with its completion. For a few, it signalled the end of their part in the SPS. For most, it heralded the moment when a readjustment would be necessary, changing from the rush of designing and modifying, progress chasing and installation, to an operational role where they became the suppliers to an exacting clientèle, and no longer members of an élite and separate corps. For some it meant assuming bigger responsibilities and the burden of the execution of other peoples designs, with which they were not completely familiar.

When the summer holidays were over, a Running-In Committee (RIC) was formed with de Raad in the chair. For him, the job of seeing that all was in readiness and deciding on the order in which the components would be brought together. His attention to detail was an invaluable asset, as from his modest office he would pose questions, or answer queries, in whatever language seemed appropriate: French, English, German, and occasionally in his native Dutch. It was for Milman and his little group to keep track of progress and coordinate the installation and testing, ensuring, for example, that sections of the tunnel could be closed off at the right periods and that everyone knew what was going on. An accelerator, however, is more than a collection of hardware that has simply to be put together correctly for it to work. Some of the characteristics can be known with the necessary accuracy only when a beam is circulating, and measurements can be made of its behaviour. Tuning a synchrotron, like tuning a piano, must be performed in a logical sequence if a harmonious result is to be achieved; here the experience of Wilson, recently back from the USA, was of capital importance. After he had seen

Europe's Giant Accelerator

*Well before installation was complete a Running-in Committee was formed under Bas de Raad (**top**) to see that the various components of the machine were brought together in a logical order and that programs were prepared for routine operational sequences. These were written by the engineers responsible for the equipment design.*

the overall design of the SPS firmly settled, and the manufacturing phase under way, he had gone off to Batavia to live through the running in of the American machine. On his return to CERN, he combined this practical knowledge with his comprehensive theoretical understanding of the SPS, acquired since its earliest days.

The presence of Adams was never far away. With Crowley-Milling as his deputy, he was there, supervising the preparations, systematically going through the essentials of what remained to be done, even after his appointment from 1 January 1976 as Executive Director-General of the combined laboratories of CERN, with a huge re-organizational task on his hands.

When the RIC went into action, much of the equipment was already in place, and either working, or approaching the end of testing. There was, nevertheless, still much to complete; but the time had come to begin to put the components together, and develop routines of operation which would bring them to life. Before the magnets could be set to develop a certain field, they had to be switched on; before that, the cooling water had to be flowing, and before that, the valves had to be opened, and the pumps switched on and so on. Small teams were assembled, each with its own leader, and assigned explicit groups of control functions. These programs had to be worked out in advance, and written by the people who had built the equipment, as only they had the specialist knowledge of why their equipment had been built as it had, and how the internal logic of the system had been conceived.

At the start of the project, a rough target date for beam tests had been set at the beginning of 1976, with a few months of slack, assuming that it might take 12 months to make the machine operational. The slack had already been taken up through slippage in the completion of the underground works while a number of plant components had been delivered late. Much of the time lost was recovered, however, by the installation teams led by Gérard Bachy, under the responsibility of Horisberger. Following the detailed schedules laid down, they were able to keep the work going smoothly even when, for example, faulty bending magnets had to be pulled out and the complex programmes had to be modified extensively.

By the turn of the year 1976, it was possible to set a definite date for the start of beam tests, and 5 April became the dead-line to which the whole organization was committed. There were still many control cables to pull through, and all the plugs to be checked to ensure they were correctly wired and there were no short circuits. The programme was tight, but feasible; although there were many late nights, there was very little panic. Not all the rectifier stations for the ring magnets would be in operation by the date set, but two would be ready, and when ring tests

One of the last pieces of equipment to be installed in the SPS ring is the heavy beam dump, which is slid into place early in April 1976. All is now ready for protons to be extracted from the PS into the long transfer lines leading to the SPS and for the SPS start-up trials to begin.

*Attention now moves to the central control room (seen from the observation tribune) as the tunnels are cleared of personnel and the auxiliary buildings, for the most part, are handed over into the sole care of the computers installed there. In addition to the three control consoles is an alarm panel (**centre**) on which are signalled any breaks in the safety interlock system or potentially hazardous situations.*

started the magnets could be powered to well beyond transition which should be enough for some time.

During the annual two-month shut-down of the PS after Christmas 1975, in addition to normal routine maintenance and the replacement of power supply cabling damaged in a fire in the previous summer, the new ejection septum magnet had been mounted in place. In gaps in the experimental programme in March 1976 the continuous transfer system, now over 10 turns, was checked out, and protons were steered with high efficiency out of the ring on to a beam dump at the beginning of the transfer tunnel. The next stage was to remove this dump, and insert one just beyond the point where the beam was due to fork away from the ISR tunnel under the Geneva–St. Genis road and on to the SPS. Again, there was no problem in getting the beam that far; the PS team's part of the contract was fulfilled.

Attention now moves to the SPS control room, a stage for drama, but not for histrionics, with a backdrop of plain beige, end walls of dark olive green, subdued overhead lighting throwing into relief the one alarm panel and three control consoles, with their gaily coloured indicator lights and television screens. Viewed from the glass-fronted tribune, which runs the length of the room on the first floor, the starkness of the walls gives quiet emphasis to the scene, where the electronic displays take precedence over the anonymous figures below. A pool of light over the log book on each desk, a built-in typewriter, a ball and a knob, controls of almost exaggerated simplicity. A single television screen on the far wall relieves the unbroken symmetry. It plots the characteristics of the PS beam that is available, and a little square of light bleeps on and off each time the beam is directed to the SPS. The operators in their chairs tap their invisible buttons, the pictures change, graphs are drawn, charts appear, a typewriter clatters, and another picture appears, a turn of the knob, and a succession of numbers flickers in their little inlays.

5 April 1976. All is in readiness for the SPS to show its paces. Without abandoning its experimenters for more than a third of the time, the PS is tuned for ejection to the SPS on alternate cycles, and a beam shoots down the tunnel in the direction of the ISR. For a few moments, nothing, as the pulsed magnet due to divert it to the SPS refuses to function. Then it comes in, and the beam monitor, placed just ahead of a beam stopper inserted before the injection magnets, 800 m away, registers practically a bull's-eye. Within an hour, the whole of the instrumentation in the transfer tunnel is checked out, the beam's characteristics are measured in four dimensions, and found to tally with the PS forecast. Already the control system is demonstrating its astonishing versatility, and refined exercises in beam manipulation are gone through. Modest satisfaction for a productive evening's work, but beam lines, even 800 m long, are almost commonplace and steering and focusing techniques well established. Meanwhile the PS returns to its routine, the beam stopper that had been installed is removed, and a clear vacuum tube is put in instead.

8 April. The experiment is repeated, this time with the injection septum magnets powered, but not the subsequent kicker magnets, so that protons should cross the SPS ring at a small angle, and finish at a beam stopper 20 metres downstream. Again, at the first pulse of the injection magnets the beam behaves as planned, and the position monitors show a direct hit on the stopper.

A succession of runs was then made during the rest of April to measure the beam characteristics throughout its trajectory, improve some of the control programs, and calibrate the automatic steering system so that only one correction would be needed to bring it to within less than 1 mm of the theoretical line. There was some gentle pulsing of the injector kicker to check its performance without pushing the beam beyond the stopper. That, too, seemed to behave. A little more quiet satisfaction, but injection magnets could be thought of as fairly standard, even if they did have to come in at a precise moment determined by the operation of another machine 1 kilometre away: anyway, the beam energy was only 10 GeV. The moment of truth, as Adams expressed it, would only come when the beam was asked to make the 7 km turn of the machine.

The date fixed for the first single turn test was Monday, 3 May, and the atmosphere was becoming tense. Had all the magnets that were faulty been replaced, or had others gone down in the meantime? Had the mounting errors detected in certain of the quadrupoles all been put right? Had the alignment teams followed their charts correctly? Had any rubbish, or an old screwdriver, been left in the vacuum chamber? (There were recollections of the laboratory which had tried putting a ferret towing a swab round its vacuum chamber to try and clean it). Were the power supplies stable? Had any of the magnets been connected up the wrong way? All preceding tests were optimistic, but only the circulation of the beam would bring proof. On the previous Thursday, the PS had been subjected to a full dress rehearsal to make sure it could deliver a fine stable beam when the time came. All was ready by the Friday afternoon. Louis Burnod, in charge of the main ring correcting magnets, and beam monitors, proposed a trial run, just to see and to give the operators a little practice before the great occasion. But de Raad was firm. Too many people had put too much into the machine to be cheated of the first moment when the first turn was tried. 10 o'clock on Monday morning was the time scheduled for the run to begin, and that was when it would be.

3 May. 10 o'clock, and the control room is beginning to fill as people drift in, some of necessity, some by right of seniority, others a little tentative, not wishing to miss the big moment, but nervous of a rebuff. Setting up the PS is signalled through, then the transfer lines, almost routine by now, switching on the main ring power supplies, setting the theoretical field, checking that all the correcting magnets are at zero, beam stopper in position in the main ring upstream of the injection point. By 11.45 there is nothing more to do, no reason to wait any longer. How far would the beam go? Through the first sextant, the second sextant, maybe the third—some accelerator builders elsewhere had taken weeks to achieve the first turn, and more weeks still to cut the beam losses to a sensible figure; thoughts of the ISR where they had done it in one go, twice, as each ring was tried—even if the ISR are only 300 m diameter, that was a bit of a precedent. The bets are laid in

The 'moment of truth' comes when the injected beam is required to make a single turn of the machine. At the first attempt, and with no correction magnets powered, the beam makes a complete circuit and, soon after, is allowed to continue circulating on each injection, for a third of a second. There is no mistaking the satisfaction that is felt by the SPS team or the fascination for the many visitors that day of the television screens which so clearly show the machine's performance.

undertones, but most people keep their expectations to themselves. Billinge breathes a confidence he possibly doesn't feel, Crowley-Milling seems abstracted, Van der Meer serious, Zettler detached (his turn would come), de Raad has too much on his plate to look anything more than anxious, Adams is non-committal, and fingers his pipe.

The 'insiders' who know the plans watch the screen on the left-hand control desk, which is connected to the television camera trained on a fluorescent screen upstream of the magnets that dump the beam when the turn is almost complete. Power the kicker—a flash—surely an illusion, the PS has stopped ejection. No, there it is again, again, and again. Stupefaction for a few moments, and then a

11/*Start-up*

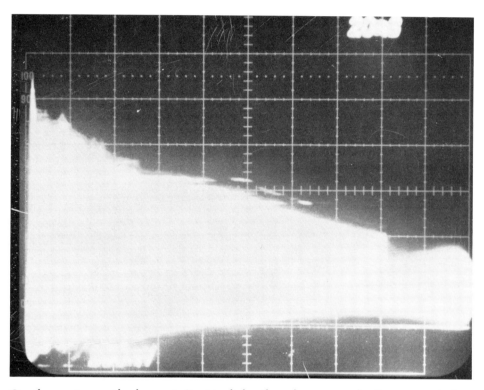

On the next run, the beam is 'trapped' by the r.f. system, the first stage before acceleration when the ribbon injected from the PS clusters into 4620 bunches, symmetrically distributed round the ring. They signal their presence by an anonymous blur on a screen.

ragged cheer as everyone crowds round to see the little circle of light pulsing regularly on and off in the middle of the screen. Not a single correction magnet powered, and there it is still in the middle. Delight mixes with relief as 7 km of near perfection displays itself for admiration.

Congratulations from the PS, congratulations all round as the control system begins to plot graphs, for all to see, of the passage of the beam around the ring. No scraping of the edges and, on average, the middle of the beam is only about 10 millimetres off-centre horizontally, and just about central vertically. Measurement of the beam losses indicates that these are less than 10 per cent from the fork in the transfer line all the way in and around to the last measurement point in straight section 6.

By lunch-time nearly all CERN knows and a steady stream of visitors stop by to gaze at the screens and murmur their own congratulations. A little celebration, but work, too. More measurements; drop the current by a fraction in the main ring magnets, and the average beam position is centred; a little fiddling with the correcting magnets, and the distance between the extreme position of the beam centre, relative to the mid-line of the vacuum chamber, all around the ring is down to 11 × 9 mm. Too good to miss. It's unscheduled, but why not take the beam stopper out, and see if the beam will go on turning for the whole of the envisaged injection period (300 milliseconds)?

Set the computer for dumping after 300 milliseconds, and there she runs for the whole of the period, another cheer. Losses seem small; take some measurements—practically no losses at all; 13 000 SPS revolutions at the first go—nearly 100 000 kilometres without appreciable loss—not bad for one afternoon.

Thursday 6 May is scheduled for the next test, when the r.f. team will take over the leading role. Problems crop up though, as the general-purpose computer (GP 3) in BA 3 is playing up; not tragic but irritating. Carrying on with closed orbit tests, the operators at the left-hand control desk tune the correction magnets and the watchers stare fascinated as the commands tap out, and the beam straightens sextant by sextant into a straight line. Over on the right, the operators are looking after the beam dumping; in the centre, the r.f. displays, but Zettler is in his own cage in BA 3. At one desk, talk is in French, at another in English, at the third a mixture of the two; a group in the centre converses in German. Senior staff look in for a moment and go back to their offices. GP 3 is working, the PS is modulating the beam; link in the r.f. and see if we can trap the pre-bunched beam. Not easy; there is a discrepancy in the frequency.

18.30 and few people are around. Zettler comes over for a chat with de Raad and Van der Meer. Abandon trying to lock on to the PS, but use a locally generated frequency standard instead.

20.00 and the control room is filling up again. Quite a few from over the road have come now; those who stayed around are regretting they didn't go and have something to eat. Schnell is there anxious to see how his baby has grown up. Surely the image there means the beam is trapped—no doubt it's nicely bunched but we're losing it. The crowd exchanges nostalgic memories of other accelerator start-ups, and how different it was in the old days. Change the Q-value. Changing the Q-value of the PS took hours of work; no question of sitting in the control room tapping a few buttons. Someone comes in to say the PS is coughing, but it seems to recover. Still no joy. Beam is unstable, better pack up for the night.

Friday evening, start again. Nasty business yesterday that earthquake near

Trieste; grunts of sympathy for the people injured and made homeless. Few have the time to speculate on what would have happened to them and their machine if the SPS had been built, as it might have been, at Doberdo, 35 km from the epicentre. Leave the PS modulator off and work from here. Beam is being captured; things look better. Acceleration for a few tenths of a second.

Capture efficiency is encouraging—about 75 per cent—and a stable and consistent situation has developed; it seems as if the phase-locking circuits are behaving as they should, and the cavities are doing what is required of them. Later in the evening, though, the beam losses are becoming troublesome; we shall not cross transition tonight. Monday is a frustrating day. The beam seems unstable and any attempt to accelerate it results in catastrophic losses at about 15 GeV after about half a second in the ring. It looks as if the time has come to stop trying to work by trial and error, and to get down to some scientific measurement.

Wednesday 12 May, and a systematic survey is started including measurement not only of the Q-values at different settings, but also how the Q-value changes with the small spread in momentum. The width of the Q-value bands, that generate resonances which result in the beam blowing up and being lost, is found to be remarkably narrow—further indication of a very clean machine, but, as expected, the spread in Q resulting from the spread in energy is big, and it is necessary to adjust the sextupole magnets to compensate the effect, and then the octupole magnets to compensate the instabilities arising from the currents induced in the walls of the vacuum tube. Nothing unexpected; the machine follows the textbooks. But now we have to pause for a week, as the PS shuts down for its regular monthly maintenance checks.

Monday 25 May, and once more we have a beam, but now we really know how it looks at injection, yet still the machine refuses to accelerate past 15 GeV. What can be wrong? Is there something in the vacuum chamber that stands up and blocks the beam when the field is raised to the equivalent of 15 GeV and then lies down again at low field? Such things had been known. Is there something funny in the control system? Is there something weird in the magnet alignment? Get Gervaise to do some re-checking. Is it the r.f. after all? This new, previously untried, system appeared to do what was expected of it, but were the sums wrong? Does it work at low energies, but break down as soon as acceleration really gets under way? Zettler is adamant. There is nothing the matter with the r.f.

As the discussions continue through the morning of 26 May an operator doggedly continues with a new series of measurement of the Q-values, made at differing times during the cycle, by kicking the beam at progressively later moments, and measuring the frequency of the oscillations that result. Everything

Acceleration though is not quite so simple, and it is soon decided that the characteristics of the beam must be studied before going on to the next stage. While the operators tap out their commands and study the responses, Wilson (left), Adams and de Raad discuss the significance of the results coming out.

is consistent with previous measurements, except near the loss point (which has been stretched to about 0·6 seconds), perhaps because low signals lead to spurious readings. Try again at midday when the PS can supply a beam.

Wilson is in the canteen about to start his dish of lasagna, when a message arrives. Can he come to the control room? Forget the lasagna; it looks as if the Q-values sag just before the loss, and the beam hits a resonance. Move the Q-values a little and try again. No doubt the vertical Q crosses a resonance point. No wonder we are not accelerating. We must change the way the quadrupoles follow the bending magnets throughout the magnet cycle so that the Q-value does not drift down as energy goes up. Modify the program and try again. Survival; two seconds circulating and 43 GeV on the screens. Transition passed without a hitch; that bogey is laid, and Zettler and his team are triumphantly

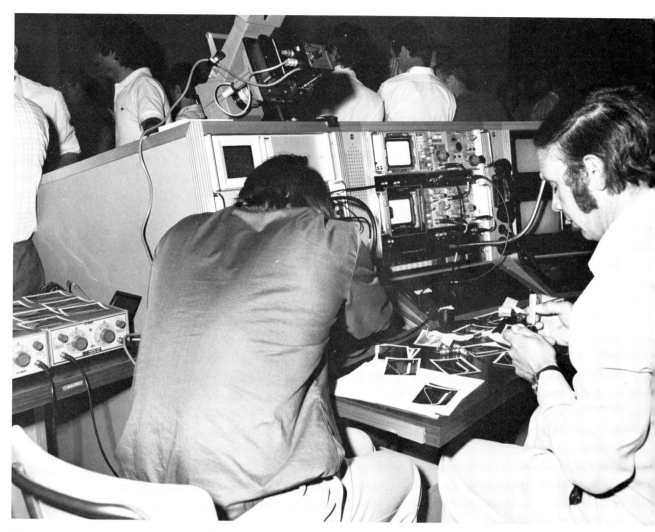

The control consoles are equipped with print-outs which can give a permanent record of the data displayed on any of the big screens. Alternatively, a photograph can be taken of the image appearing on the smaller screens. Here Wilson is sorting through the latest batch of pictures, that show that the maximum energy attainable with the power supplies then coupled in is being systematically reached.

vindicated. He is not surprised! A bit more work on swinging the Q adjustment, and 80 GeV is registered, the limit to which the two power supplies connected can take us. The SPS after only eight days of testing with a beam, is the highest energy accelerator in Europe.

Now it is up to Van der Meer and his too-small team working all the hours they are awake to bring in the other power supplies. By 4 *June* six of the twelve are coupled in and the target is 200 GeV. The beam is beautifully central in the vacuum tube, but still some losses after half a second or so. However, there is more experience now in computing the adjustments necessary to the Q-values. Feed these in, and 200 GeV it is, with no losses during the entire acceleration cycle after bunching.

All eyes are now on the calendar. On June 17 at 2.30 p.m. the delegates from member states would begin their half-yearly Council session to hear reports of progress from the Directors-General, and consider the budget for the following years. Could they make it in time? No commitments had been made, but the delegates had been informed of the successful first beam tests, and no-one in the SPS was in any doubt that impressing the delegates was of capital importance.

More supplies are hooked in: 280 GeV, and now instabilities appear as the magnet current rises. The whole situation is enormously complex, with so many individual circuits interacting with each other; there are only so many hours in the day, and tracing the causes, modifying time constants in the six buildings strung out round the 7 km ring, four in France and two in Switzerland, all takes time. Van der Meer, working a 16-hour day, impervious to tiredness or nerves, ploughing his way steadily through. But there are no short cuts, and those not involved condition themselves to disappointment.

17 June: while the Committee of Council prepares for the plenary session in the afternoon, the SPS gets ready for another attempt. Mid-day, and all twelve rectifier stations are coupled up and seem to be behaving. Try for 300 GeV, the original programme target—carefully, systematically, calculate the Q off-set, and set the controls. 300 GeV it is; quite big losses, but no time to fiddle. A note is sent to Adams.

A little after the scheduled meeting hour, the delegates take their seats, a crowd of CERN senior staff in attendance, the preparatory business is gone through, the agenda agreed, and Adams begins his report. No bursting with the news; a dry factual account of the steps that have been taken during the reorganization of the laboratory, the work on the synchrocyclotron, the experimental programme of the PS, the performance of the ISR, leading up to a measured statement on what has been happening at the SPS since December last, finishing with the bald announcement that that morning the SPS has attained 300 GeV, the maximum operating energy formally approved by Council. Gravely, Adams reminds Council that because of the Dutch reservation, operation at higher energies has not been sanctioned, but he hopes that Council would allow them now to go higher. Council agrees, and it is time for tea-break. A note is passed to Adams, but delegates are already on their feet and moving out of the Council chamber. Twenty minutes later when the delegates reassemble, having noticed that Adams had not been available during the break for the usual informal chats, Adams announces that subsequent to the Council's approval the SPS had operated at the maximum design energy of 400 GeV. The chairman expresses Council's satisfaction; next item of business.

11/Start-up

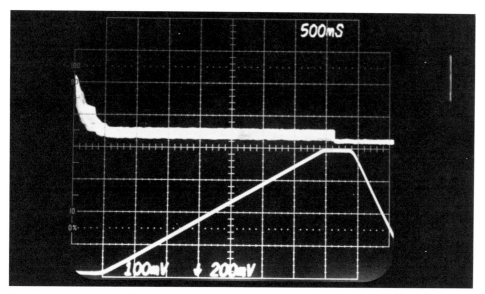

On 17 June 1976, the great moment comes when, with all the power supplies running stably together, protons are accelerated to the maximum energy of 400 GeV, four seconds after injection. At this stage, the losses during the first half second are significant, but the reasons are soon discovered. During the commissioning of the extraction systems and beam lines, the machine is run with a low intensity, but in October, full intensity beams of 10^{13} protons are being accelerated to maximum energy as a matter of routine.

But even Councils relax. Formal business over, and everyone troops over to the SPS to see the new wonder, admire the control room, believe that what they see is indeed a 400 GeV beam of protons, drink a little champagne, and hear Adams record, on behalf of all, the splendid success of this European collaboration, and the men and women who had made it possible. A simple party, but someone had had the foresight and the confidence to prepare it.

Not quite the end of the story. For Lévy-Mandel, handing over to Horisberger while taking up an appointment to join Wüster as a director of the combined laboratories; for Crowley-Milling, the SPS division to look after; for Brianti, half the job lies ahead; and for Gervaise, Van der Meer, and Göbel (now responsible for radiation protection of the whole of CERN), much to do in making the north site ready for experiments. For Zettler, the r.f. to be tuned to its ultimate efficiency and dreams of new systems; for Billinge and Wilson, some long-term thinking on what might happen next; and for Klein, a return to the French civil service.

For de Raad, the responsibility of getting the machine into a fully operational state. The beam losses have still to be sorted out, they now seem to be due to an uneven shifting of the Q-values with energy, in spite of the new programming of the quadrupoles in relation to the bending magnets. An inspiration, the response times of the measuring coils in the control magnets are checked and found to be different; only a few hundredths of a second lag between them, but enough to make the Q-values droop. The answer is to compensate the time constant.

The extraction system must now be tested, once the current period of thunderstorms is over. Lightning striking as far away as 100 km causes the main breakers on the grid line to snap out and in, and to shut down the computers. Then the memories have to be reloaded, but another lightning strike follows, and they are down again. The solution will come in the autumn when the computers will be fed through accumulators.

By the end of July there has been one good run with the ejection system to the West area. All the real fences have been cleared, the SPS is now in the home straight, heading for 10, 20, how many years of operation? The PS has already completed 17!

What of costs? It is a little early to make the final estimate, but they will be within the sums agreed.

12/A Tool to Use

Fascinating though the construction of an accelerator may be, and however commercially useful the technologies that are developed in the process, the SPS is still essentially a tool; a tool for doing experiments in particle physics. From the beginning, this has been in the forefront of the minds of the builders, and the delegates of the member states financing this great European enterprise. Even before the main characteristics of the machine were decided upon, teams of European physicists, under the auspices of the European Committee for Future Accelerators, had studied the sort of experiments that should be carried out, and in 1967 had produced a massive three-volume report on those which at that time seemed to be most relevant. In the estimates of costs of both the early machine, and then the SPS, a proper figure was included for the construction of experimental halls and beam lines, and due warning was given of the cost of building up experiments to exploit the facilities provided. Of the agreed ceiling of 1150 million Sw. Fr., the machine alone was estimated to cost 750 million. Of the remainder, 100 million was allocated to operational costs during the programme period and 160 million to the construction of the North area. To re-equip the West area for experiments at the new energies, funds had to be found elsewhere in CERN's budgets. The capital cost of the three large detectors installed there, and originally built for experiments at lower energies, already amounted to about 150 million Sw. Fr., but a big reconstruction was necessary to be able to take advantage of the higher energies that would become available, and two main ejection channels that would serve the area would have to be built. It was possible to save nearly 20 million Sw. Fr. by re-using beam line material from the redundant ISR–west area link, but altogether over 150 million Sw. Fr. more had to be found by CERN to prepare the area for the new research.

The Working Group on the Experimental Areas was the largest of those associated with the Machine Committee, and included Adams himself. In it, the essential features of the areas were hammered out. In 1973, an SPS Experiments Committee was formed, alongside the three other CERN Experiments Committees, to study applications from physicists in the member states to do research with the

12/A Tool to Use

Giorgio Brianti, already many years at CERN, during which time he had led the Synchro-Cyclotron Division and the group which built the Booster, was chosen to coordinate the building and equipping of the experimental areas, a task needing diplomatic skill as well as scientific experience.

Opposite

The SPS is equipped with two systems for extracting the accelerated protons out of the machine and leading them towards targets in which new particles (hadrons) are produced for serving the experimental areas. The first area to be equipped was the West zone, and experiments started there while the North Area was still under construction.

One of the most important programmes of research is the study of the interactions of neutrinos with matter. Neutrinos (and muons) are produced when hadrons, created when protons strike a target, decay in a 300 m long evacuated tunnel. To help define the neutrino energy, the secondary particles which have not already decayed are absorbed in an iron quadrant. The SPS generates the most intense and defined neutrino beams in the world.

new accelerator, maintaining continuing contact with the building programme. The man chosen to head the Experimental Areas Group within the SPS team was Giorgio Brianti, a red-headed Italian who had led the Synchro-Cyclotron division at CERN over many years, and had taken over the building of the PS 800 MeV Booster as the difficult SC re-build was nearing its conclusion. Known not only as an excellent physicist and engineer, but also as the best division head in CERN from an administrative point of view, his became the delicate task of converting all the demands of the experimental physicists into practical solutions compatible with the machine's possibilities and the budget allocations. His technical experience, administrative skills, and invariable good humour would all be needed if

12/A Tool to Use

Amongst the detectors used to study neutrino effects is a 20 m long assembly of magnetized iron disks, interspersed with scintillation counters which are designed to record as far as possible all the interaction products arising from the collision of a neutrino with another particle. Very few interact conveniently near the beginning of the chamber; only 10 per cent of a beam of high energy neutrinos passing through the Earth's diameter would, on average, collide with the matter in its path.

the large capital expenditure on the accelerator was to be exploited to the full. The task of constructing the ejection tunnels and the buildings in the North experimental area fell once again on the broad shoulders of Lévy-Mandel, although later they became the responsibility of Hans Horisberger.

High energy physics had advanced a great deal in the years since the 1967 ECFA study, and much of the detail worked out then had become out-dated. Moreover, it was both economic as well as expeditious to design the West experimental area around the buildings and big detectors that existed. Happily, it was also good physics. The main hall, that once had been used as an assembly place for the ISR, and then converted to PS physics, was suitable for experiments with secondary beams of energy up to 200 GeV; it contained the versatile spectrometer Omega in one corner, and led naturally to the 3·7 metre hydrogen bubble chamber BEBC. Behind BEBC, there was room for other big detectors suitable for doing experiments with neutrinos up to the full energy possible. Neutrinos have been regarded for many years as prime particles for exploratory work. Like photons, the particles of light, they have no mass, but, unlike them, they react with matter only through the weakest of the four fundamental forces; in consequence, they make a pure probe for investigating this force. On the other hand, the chance of one colliding with another particle within any detector is very low, but it does rise at higher energies. Interest in neutrino physics has greatly increased over the past years, in particular through the experiments with CERN's heavy liquid bubble chamber Gargamelle, which it was decided should be dismantled and rebuilt behind BEBC in line with the neutrino beam. Between these two huge bubble chambers, a massive assembly of counters has also been built so that the neutrino beam can serve all three at once. The position of the West Hall, and BEBC in particular, was one of the main factors determining the exact siting of the SPS, and the point at which the proton beam should be extracted from straight section 6.

The neutrino beam line is one of the most bizarre features of the experimental facilities, as over the first few hundred metres particles are travelling in vacuo, and over the next few hundred through 6000 tons of iron, then earth, and the foundations of the West Hall. After being ejected from the main ring, the protons are led into a tunnel section, 170 m long and 6 m diameter, where they strike one of two targets of beryllium, a metal chosen for its lightness and good heat conductivity (aluminium can also be used).

Typically 3 millimetres diameter, the complete target consists of between 5 and 11 rods, 10 cm long, set end to end and supported by thin beryllium discs. The thermal shock on the target with a fast ejected beam is so high that only small lengths carried on light supports and cooled by helium gas will survive. In the target, a spray of secondary particles in the hadron family is produced: these are the parents of the neutrinos, also a potent source of radiation. As a shield against

Opposite

In line with the neutrino beam are two massive bubble chambers. BEBC (shown here) *contains a cylinder, 3·7 m in diameter, filled with liquid hydrogen, or a liquid hydrogen–neon mixture. The chamber body is surrounded by a superconducting magnet, enclosed within an iron shield. Photographs taken in the bubble chambers are distributed to many laboratories for analysis.*

12/A Tool to Use

this radiation, the targets and attendant beam monitors are enclosed in an iron box giving a minimum thickness on the access side of 80 cm. Even following a day's 'cooling', the radiation dose in the region after a machine run will only allow people to approach for short periods, while with the beam in operation, the radiation levels locally will be lethal. As a result, all fine adjustments of the target have to be made remotely. Although from the protons' view-point the beryllium rods are a target, for the neutrinos they are the guns firing at a bull's-eye 5 cm across and 1 km distant. Precision and reproducibility in adjusting the rods in three dimensions is crucial to providing an intense neutrino beam of known characteristics.

On leaving the targets, the hadrons go either through two focusing horns for the 'wide band' beam, or through a collimator and a system of magnetic analysing magnets and focusing magnets for the 'narrow band' beam. The band refers to the range of neutrino energies which will arrive at the experiments, and can be limited by first of all selecting the energy of the parent secondary particles which are aimed at the experimental area. For wide band operation, the maximum number of neutrinos possible is desired; to focus as much of the spray as possible into a parallel beam, the emerging secondary beam is directed through two horn-shaped deflectors, set 2 m and 80 m downstream from the target centre, and which carry a pulsed current of several hundred thousand amperes.

On to the next section, which is a pipe 300 m long and 1·2 m diameter, evacuated down to low pressure by water turbines mounted on the surface. The jet of water produced by the turbines entrains the air in the pipe, and passes it on to rotary backing pumps. While travelling down the pipe, the hadrons decay, each producing a neutrino and a muon, which continue more or less along the same line of flight. The pipe is evacuated to prevent scattering from air molecules the hadrons would meet en route. Paradoxically, provision is made for moving into the line of the narrow band beam a 3-m long quadrant of iron. This is to stop those hadrons which have not already decayed near the beginning from continuing down the tunnel and producing neutrinos, which make a bigger angle with the centre line, hence have a different energy, but still enter the detector. Closure of the pipe at the upstream end where the window has to be thin, again because of scattering, presents a nice safety problem, because if it were pierced when anyone was in the connecting tunnels, the suction would be irresistible. The damage to equipment would also be serious. The window is made of 2 mm thick titanium, and must be covered by a thick iron plate when people are working nearby. After the decay tunnel comes the 180 m absorption tunnel, where the remaining secondaries are quickly absorbed and the unwanted muons are gradually slowed down. Inside this

12/*A Tool to Use*

tunnel are 425 iron disks, 40 cm thick, surrounded by weak concrete, which, it is hoped, will break up easily when the time comes to recover the iron. At periodic intervals, gaps have been left for muon detectors, which are accessible from a service tunnel running at the side. Measurement of the intensity and energy of the muon flux along the length gives additional information on the neutrino spectrum. BEBC is reached through another 300 metres or so of earth foundations and walls, and beyond BEBC are the counter array and Gargamelle.

BEBC is one of the largest low-temperature bubble chambers in the world. Built under a tripartite agreement between France, Germany, and CERN, it encloses about 35 000 litres of liquid gas in a vertical domed cylinder, 3 metres high and 3·7 metres diameter. The body of the chamber is surrounded by two superconducting coils, cooled to liquid helium temperatures, which produce a magnetic field in the body almost twice that produced in the SPS bending magnets. Expansion and recompression of the hydrogen is produced by the sudden movement of a piston moving in a cylinder below the chamber, which leaves the top of the chamber free for the lights and cameras which photograph particle tracks formed in the liquid. The assembly is enclosed in a massive iron shield to contain the magnetic field.

It is calculated that at full beam energy and intensity, with the wide band beam about 50 thousand million neutrinos will arrive at BEBC for every pulse of the SPS. Out of these, the number of neutrinos which make collisions in the chamber will be less than 0·3 per ton, an intriguing unit which implies that when filled with a neon–hydrogen mixture which has a density 10 times that of hydrogen, every picture should show about five interesting reactions. If filled with hydrogen, one out of two pictures should be interesting. With the narrow band beam the event rate drops by a factor of 10 to 100. Modest though these last figures are, they are a great improvement on what had been possible before.

To begin with, the chamber will be operated with a neon–hydrogen filling and a narrow band beam, but afterwards an inner chamber will be installed containing pure hydrogen to provide a simple target, while retaining the higher absorption qualities of the mixture to detect the reaction products.

Gargamelle also, is one of the largest bubble chambers of its type. Its body is a horizontal cylinder, 4·8 m long and 1·88 m diameter, which can be filled with either the fuel propane, or the liquified gas freon, that is used as a pressurizer in modern sprays. Pressure control is through a series of quick-acting valves, which connect high or low pressure tanks of nitrogen to manifolds closed on the chamber side by flexible membranes. The iron core and magnet coils surround the cylinder in the axial direction, leaving room on one side for the flash lights, and on the other for eight cameras which photograph the interior. Weighing more than 1000 tons, the chamber has been mounted on a pedestal, 11 metres high, with 6·5 metres above ground to bring it into line with the neutrino beam. The shock to the foundations on the compression and decompression strokes of both Gargamelle and BEBC is impressive.

The film from the bubble chambers can be so readily distributed to other laboratories for analysis that it is often difficult to say exactly who will be participating in the experiments, but a year before Gargamelle was due to go into action, three experiments had been approved—collaborations of Bari, Milan, Orsay, Palaiseau and CERN; Aachen, Bergen, Brussels, Strasbourg, and University College, London; Orsay, Strasbourg, Ecole Polytechnique, and CERN.

Between these two chambers has been installed an electronics experiment, almost equally imposing in its dimensions. The main feature is a 20 metre line of magnetized iron disks, 3·75 metres in diameter, interspersed with plates of scintillating material viewed by high sensitivity electronic detectors. The total mass of the detector is around 1400 tons. This experiment is a collaboration of Dortmund, Heidelberg, Saclay, and CERN.

Quite apart from the neutrino beams, two other beam lines rise up from the SPS in a separate tunnel towards the West Hall. In one, a target is installed early on to give room for analysing and selecting with great precision hadrons destined for BEBC. From the other line of protons, up to five beams can be produced of different types from targets mounted at the surface. The most unexpected is the high-energy electron beam that is produced from neutral hadrons, the majority of which decay into a gamma ray and an electron–positron pair, giving rise to a beam containing up to 10 million electrons at 80 GeV. When these strike a sheet of lead, gamma rays are again produced as the electrons slow down, and measurement of the final electron energy allows the gamma ray energy to be calculated. These gamma rays (photons) of energy up to 70 GeV, then strike a hydrogen target, where, it is hoped, they will produce evidence of charmed particles.

Searching for 'charm' is the second most popular pastime at the beginning of the SPS programme, but rather than go into the detail of all the experiments planned and approved, let us simply look at the list of participating universities and institutes:

Bristol—Heidelberg—Geneva—Orsay—Rutherford—Strasbourg
Amsterdam—CERN—Cracow—Munich—Oxford—Rutherford
Bonn—CERN—Daresbury—DESY—Ecole Polytechnique—Glasgow—
    Lancaster—Manchester—Orsay—Sheffield
Indiana—Saclay

Looking back along the lines feeding the West experimental area. In one (left) hadrons are produced early in the line and the remaining length is used to select with great precision, the required type and energy. The other, feeds a number of target stations on the surface, from which a series of beam lines fan out, serving the various electronics experiments.

CERN—Trieste—Vienna
CERN—Genoa—Orsay—Oslo—University College London
Birmingham
Clermont Ferrand—Leningrad—Lyon—Uppsala
Geneva—Lausanne
Birmingham—CERN—Neuchâtel—Rutherford.

In the North area, the emphasis will be on experiments with hadrons up to the highest energies and on the so-little-understood muons. All experiments to date have failed to discern any feature which distinguishes the muon from the electron, apart from its mass and its life-time, which is about two millionths of a second.

The beam lines, separated by concrete blocks, contain magnets and electrostatic or r.f. devices for guiding the particles and selecting the type and energy required. Fast-acting switching magnets can bring more than one target into use during each SPS pulse, and a single target can supply a number of experiments with different particles.

In one corner of the West Hall is a big versatile electronics spectrometer. Inside a large volume superconducting magnet, an array of spark chambers can be rolled into place. The spark patterns are recorded from above through television-type cameras as electronic signals, which are analysed by a computer. The events can later be reconstructed for visual study if required.

On levelling out, a few metres below ground level, the beam of protons directed towards the North Area is split and then led to a domed building (now covered with earth) in which the targets are situated. The building is massively shielded with concrete but, as a precaution, is equipped with a special drainage facility to ensure that no radioactive water is allowed to flow into the local streams.

Rising up from straight section 2 in the SPS ring is an ejected proton beam line, which flattens out after 600 m about 10 metres below ground, and then enters the splitter tunnel, where it can be divided into three in two stages. From there, the beams move into the target area, which is an underground hall, 130 m long by 16 m wide, capped by a domed roof 10 m high in the centre, and covered with 10 m of earth. This is potentially the most radioactive zone in the north area, and special safety features have been provided. The walls and floor are particularly thick, and drainage water is collected into a sump, where it can be checked before being pumped to the normal drainage system. In the hall, two of the targets can each provide two hadron beams, which are then led away in pairs up to the first experimental hall 270 m away, while the other target provides the muon beam through the decay of hadrons along a 480 m long transfer tunnel. In addition, a proton beam can be led along this tunnel and into what is at present a blind stub, but which can be extended in the future when the need arises.

The bends in the hadron lines can be used for analysing the beams on their way to the experimental hall, which is 290 m long by 50 m wide, and which gives a clear height of 10 m under the hooks of the two 40 ton cranes. Here, electronics experiments will be laid out similar to those in the West area, but more elaborate. When the muon beam leaves the proton line, it rises upwards, and after 500 m flattens off to enter the second experimental hall. Throughout its run, it is focused, while the bends in the line are used (as for the hadrons) to define the

energy of the particles accepted. The second hall, 98 m long by 17·8 m wide and height similar to that of the first hall, is built in a natural hillock. It has been re-covered with earth to restore the original contours. Tunnels link it to the access road which serves the area. Great care has been taken in the design of the North area, not only to make it as inconspicuous as possible and to preserve the natural woodland, but also to direct the beams in such a way that most of the stray radiation is absorbed in the earth.

So many people wished to participate in the first muon experiments, that individual institutes and universities are no longer cited, and it is called simply the European muon collaboration. Some 60 people are listed and they come from all the corners of Europe. In the hadron experiments already approved by the summer of 1976, the lists are more manageable:

Frascati—Milan—Pisa—Rome
CERN—Collège de France—Ecole Polytechnique—Orsay—Saclay
CERN—Dubna—Munich—Rome—Saclay
Max Planck Institute, Munich.

As the experiments become more complex and more expensive, the number in the team tends to rise, and it is not unusual now to find 20 and more names figuring against each experiment. The number of proposals for experiments has also played a part, as it would have been impossible to accommodate everyone if the size of the groups had not expanded. This has meant that experiments tend more and more to have a number of different objectives, and the equipment is designed with more flexibility than was common in the past. Even before the SPS started, there were probably over 400 physicists in western Europe making active preparations for the research programme. In addition, there is participation by US universities, attracted by the greater precision and reliability of the European machine, and east European countries have also expressed their desire to join in.

The accusation has often been made that when the PS, the world's first strong-focusing proton synchrotron came into operation, the physicists were not ready to do experiments with it. The same could not be said when the world's largest machine, the SPS, started up.

Opposite

In line with the splitter and target buildings is the large hall in which experiments will be made with hadrons of the highest energy. Bearing away to the right and beyond is the muon line, ending in the hall where experiments with the most intense and closely defined muon beam in the world will be done. These might answer the 40-year old question: Why did Nature need the muon?

Europe's Giant Accelerator

Dr. John B. Adams, Executive Director-General, CERN

13/Epilogue: the long search continues

Will the machine builders now rest on their laurels and be content to make marginal improvements to what they have? It seems unlikely. Studies have been started already on the possibility of adding storage rings to the SPS. One possibility would be to replace the conventional magnets in the ISR with superconducting magnets, to enable colliding beams of protons of 100 GeV to circulate, but when 400 GeV is available it seems more tempting to think in terms of a new ring concentric with the SPS ring, where 400–400 GeV reactions could be studied.

An exciting thought is to build a storage ring for anti-protons produced by the PS which could then be injected into the SPS already filled with protons of the same energy travelling in the opposite direction. The two beams would then be accelerated together to higher energies. There have been semi-serious discussions about a world machine for 10 000 GeV protons or more. But the political and organizational problems are such that the present feeling is that the continental scale, or, in Europe split in two, the semi-continental scale, is still the one to consider.

On the other hand, perhaps the next generation will be more concerned with electrons. In Hamburg, 19 GeV electron–positron storage rings are under construction, and while a big international participation is expected, this has not prevented ideas being formulated for a European machine which would allow experiments to be made with colliding electron–positron beams, each of 100 GeV, in a ring 5–6 kilometres in diameter.

In most of these schemes the order of cost would be equivalent, at least, to two SPS: this is hardly an expenditure that many European states would be prepared to contemplate for a long time to come. High energy physics can anticipate at best a constant budget for many years, out of which the present machines will have to be run. It is conceivable that when the experimental programme on the ISR begins to run down in, say, five years' time, it would be wise to shut it down and transfer the funds saved to a new project. By vigorous pruning of the less interesting physics perhaps 100–150 million Sw. Fr. could be allocated to a new European constructional programme, which would suggest a minimum of 10 years for its completion. This is already a long time. The SPS came into operation nearly 15 years

Europe's Giant Accelerator

after first discussions started, although the detailed design and construction of the machine took less than six years. If we extrapolate this time-scale to an enterprise double the size, it risks exceeding the active working span of a person's life. This may be the limit to his aspirations, as planning machines for one's children may not be very attractive, and it is more than probable that the children would not appreciate them when they were finished. Is the SPS then as big as we should go? Let us see what new physics it brings before attempting to answer.

Appendix I/Particle Accelerators*

Nuclear particle accelerators, although necessary for research into the fundamental structure of matter, are only a means to this end.

They started from very humble origins as instruments which could conveniently be placed on an ordinary laboratory bench. They were built by the physicists carrying out the research as a part-time activity. Today, an accelerator is built by an expert team of engineers and scientists who will not, in general, engage in the research which the accelerator makes possible. Whereas 40 years ago an accelerator could be built for a few thousand pounds in a university laboratory, the most recent machines cost many millions of pounds, and require national and international laboratories to be established for their construction and operation.

All this has happened over a period of about four decades, during which very great changes have taken place in experimental apparatus, in research method and style, and in the organization of research, resulting from the development of the giant machines.

The first accelerators were built and used for nuclear research in the early 1930s. Before that time, only natural sources of nuclear particles were available: these were the naturally radioactive elements and cosmic rays. Rutherford in his research work in 1919, which led to the discovery of the atomic nucleus, used alpha particles from radium or thorium sources. The energy of such particles is between 5 and 8 MeV, and it was not until about 1935, that particle accelerators could reach these energies.

Thus, the first particle accelerators were developed with the aim of replacing the natural radioactive sources by machines which could produce beams of particles, of similar or higher energies. The research which could be done with such machines was already tentatively explored using the natural radioactive sources. This pattern of exploration with natural sources, followed by detailed studies using particle accelerators runs throughout the history of this work.

---

* This text is based by agreement with Dr. Adams, on the 4th Walther Bothe Memorial Lecture, 1971, given at the Max-Planck-Institut für Kernphysik, Heidelberg.

Europe's Giant Accelerator

Cecil Powell (in his Walther Bothe memorial lecture in 1969) spoke about the nature of the primary cosmic radiation, and traced the course of this research from the observation by Coulomb in 1785 that the electric charge on a metal sphere suspended by a long silk thread slowly leaks away, to the demonstration by Bothe and Kohlhörster in 1929 that part of this leakage was caused by penetrating material particles coming from outer space. These cosmic rays provided the nuclear physicist with a weak, but very high energy, source of particles for the experimental study of nuclear reactions.

Thus, even before particle accelerators had been developed sufficiently to replace the natural radioactive sources, another much more formidable challenge was presented to the accelerator builders—to replace cosmic rays by man-made machines. From about the 1940s onwards it has been primarily this challenge which has inspired accelerator development. Only in recent years have the accelerator builders seen a way of producing in the laboratory nuclear events comparable in reaction energy to those produced by the incoming cosmic rays.

Not all physicists have welcomed the success of accelerator development. There were certain cosmic-ray physicists who found real physical and mental satisfaction in working in laboratories on top of the highest mountains, and in constructing and launching large balloons. Curiously enough, the mountains and launching sites always seemed to be selected in very pleasant parts of Europe. Powell in his lectures deliberately used to annoy the accelerator users by producing from his pocket a small tin of photographic plates, and explaining, with becoming modesty, that this was all he needed as experimental apparatus—the rest was supplied by nature and could be obtained by a healthy climb up a high mountain. Of course, he was only teasing. Nobody was stronger or more active than he in supporting the building of accelerator laboratories in Europe when it became clear that these were necessary for the continuation of the research.

Early ideas of machines to accelerate nuclear particles all took the form of a high voltage source, coupled to a long evacuated tube made of an insulating material, with the particle source at the high voltage end, and the target to be bombarded at the zero potential end. The diversity in these early machines lay in the various types of high-voltage sources employed, and in the ways of making and evacuating the accelerating tubes.

Around 1900, the most popular high voltage source was the induction coil, generally known as the Ruhmkorff coil, which was capable of producing a few hundred thousand volts. It consisted of a core of straight iron wires wound with a few layers of thick insulated wire, called the primary coil, on top of which were wound thousands of turns of fine insulated wire to make a secondary coil. An automatic switch made and broke the current in the primary, and induced a high potential across the secondary. The accelerating tube in those days was usually made of glass, for example the Röntgen or X-ray tubes, and the air from this tube was evacuated either by Sprengel or Toepler mercury pumps and, later on, by the Geryk cylinder pump. Electric currents were supplied by galvanic cells, or by accumulators.

An advertisement in 1897 by an English firm for this sort of equipment is shown overleaf. Some idea of the attitude of experimenters towards their equipment can be learned from a leading article in a scientific journal of 1898 which said: "The possessor of a good induction coil made by our leading instrument-maker should cherish it as a violin player cherishes his Stradivarius".

The induction coil gave pulses of high voltage at the frequency of the make and break contact, but there also existed machines giving continuous high voltage. These were the separation of charge machines, called the Toepler-Holtz or Wimshurst machines, invented between 1870 and 1890. Such machines, whose operation is quite complicated, using counter rotating discs 40–50 cm in diameter, could develop about 100 kV, and deliver currents of several milliamperes.

Both the induction coil and the separation of charge machines were developed in the early 1930s to act as high-voltage sources for nuclear experiments. For example, a variant of the induction coil, called the Tesla coil, was developed in 1930 to give about 1 MV, and a Holtz type generator built by Dahl in 1935 gave a current of 10 mA at 200 kV. The Wimshurst machine has been subject to recent development, in the form of dielectric cylinders rotating in a hydrogen atmosphere, by the French firm of SAMES, who now produce very reliable machines giving d.c. currents of a few milliamperes at voltages between 50 kV and 1 MV. These machines are often used to preaccelerate protons for injection into the most modern and largest nuclear particle accelerators.

Another important offspring of the separation of charge machines was the modern electrostatic generator, using an endless belt to transfer the electric charge, whose development was started in 1930 by Van de Graaff. The voltage of the first machine built was about 1 MV. Two recent machines of this type, one at MIT and the other at Los Alamos, have achieved 9 and 8 MV respectively and tandem versions operate at about 20 MV effectively, by using the applied voltage twice.

The other early types of high-voltage generators aimed at the multiplication of voltage in stages to reach higher voltage levels. For example, the surge generators using the 'Marx' circuit stored electrical energy in capacitors charged in parallel, and then discharged this energy by means of spark gaps which connected the

Advts.]

Apparatus for

..."X RAYS" PHOTOGRAPHY.

FOCUS TUBES,
25/- each.
Each tube is tested before being sent out, and will produce brilliant negatives with the shortest exposures

INDUCTION COILS,
From £8 10s.

FLUORESCENT SCREENS,
Clearly showing the Bones of Hand, Arm, etc

BICHROMATE BATTERIES, ACCUMULATORS.

PRICE LIST POST-FREE ON APPLICATION.

☞ The Apparatus can be Seen and Tested at our Show Rooms.

G. HOUGHTON & SON,
89, HIGH HOLBORN, W.C., 1897

Telegrams "Bromide, London."

Appendix I/Particle Accelerators

capacitors in series. One of the highest voltage surge generators was built by General Electric at Pittsfield in 1932 and produced surges of over 6 MV. The pulse duration of these machines is, however, only a few microseconds, during which time they are capable of delivering currents of the order of 1000 A.

Many other ingenious circuits were invented and built to achieve voltage multiplication, such as the cascade transformer machines and the transformer rectifiers, but the most well-known system, which by common consent is usually regarded as the first nuclear particle accelerator used for research, was the voltage multiplier built by Cockcroft and Walton at Cambridge University in 1932, and used by them in a series of experimental studies of nuclear reactions.

About £10 000 was spent building this machine. The first machine gave about 500 kV, but later Cockcroft and Walton installed a 1·25 MV voltage multiplier which was engineered and built by the Philips Company at Eindhoven. This firm has specialized in the construction of these machines, and many of the largest particle accelerators now operating in the world use Philips machines of about 500 kV potential as pre-accelerators in their injection systems.

All the early accelerators, whatever type of voltage source they used, involved the development of evacuated accelerating tubes, and this raised a number of technical problems very difficult to solve in those days. The accelerating tube for instance, had to be divided into sections to withstand the high voltages without electrical breakdown. Each section consisted of an insulating cylinder, usually glass or porcelain, and a hollow metal electrode through which the particle beam could pass.

Thus, the voltage was divided down the length of the accelerating tube, and by carefully shaping of the metal electrodes the beam of particles could be focused electrostatically during acceleration. Cockcroft and Walton, for example, used glass cylinders made for the petrol pumps common in those days, and stuck these to the metal electrodes with plasticine.

Later on, techniques became more sophisticated as glass-to-metal seals were developed and oil and mercury diffusion pumps became available. Electrical breakdown, which originally occurred inside the accelerating tube due to poor vacuum and contamination, was transferred to the outside of the tube, and, in place of air outside, it became common to use various gases, such as freon or sulphur hexafluoride at relatively high pressures.

But, despite all the technological improvements, it became clear, even in the 1930s, that the way to higher particle energies was unlikely to be through the development of higher voltage sources and longer accelerator tubes; this has proved to be the case. Such devices can accelerate particles to a few tens of

MeV, but for higher energies entirely different machines were to prove far more effective.

At the time that Cockcroft and Walton were building their accelerator, Ernest Lawrence, at the University of California, was developing an entirely new principle of accelerating particles—the principle of resonant acceleration—and building a new type of machine which he called the cyclotron, incorporating this principle.

Lawrence proposed the cyclotron principle in 1930, having conceived the idea in the summer of 1929 while looking through a paper which Wideröe had published in 1928 in the *Archiv für Elektrotechnik*. The idea, simply, was to circulate the particles round and round in a magnetic field so that on each revolution they could be accelerated by an electric field set up at two points on their circumference. Thus, the same electric field would be used time and time again as the particles gained energy, and the difficulties of high-voltage breakdown in accelerator tubes would be avoided.

Particles in a cyclotron start from a source at the centre of a magnetic field, and steadily gain energy on each revolution as they traverse the electric field: as they do so they spiral outwards. At each turn of the spiral, the radius of curvature of the charged particle is such that its centrifugal force exactly balances the magnetic force. The important point is that, in a time-constant magnetic field, the frequency of revolution of the particle is constant, independent of the energy of the particle, so long as the particle mass remains constant.

The magnetic field for a cyclotron is provided by a large d.c. magnet with circular poles, and a gap between them in which the charged particles circulate. An ion source is mounted at the centre of the cylindrical magnetic field, and the electric accelerating field is established tangential to the particle orbits by hollow electrodes, called Dees. The particles cross the electric field twice per revolution and, since at each crossing the electric field must be in the correct direction to accelerate the particles, an alternating electric field is required. For protons, the frequency of the electric field for a magnetic field of 13 kG is about 20 MHz. Thus, a high frequency alternating electric field is required for acceleration.

Since the particles circulate many hundreds of revolutions, their beam must be focused during their acceleration either by shaping the electric field or the magnetic field. In the latter case, which is the most usual, a slight gradient of the magnetic field gives the required focusing forces; it is customary to shim the magnetic field guiding the particles round their circular orbits, so that the field decreases slightly from a maximum value at the centre of the magnet to a value a few per cent lower at the periphery of the pole tips. The curvature of the field lines due to this negative field gradient gives radial field components which return the circulating particles to the median plane of the magnetic field.

Cyclotrons of ever-increasing size and particle energy were built from 1932 up to the outbreak of the Second World War. The Lawrence Radiation Laboratory (as it is now called) at the University of California, Berkeley, was the principal centre. The team of physicists there became world-famous, and a new breed of physicists—the accelerator physicists—evolved.

By 1937, it was recognized that cyclotrons using the resonance principle were limited inherently in the maximum energy to which they could accelerate particles. This limitation arises because the mass of an accelerating particle does not remain constant at its rest mass, but increases with its energy, according to relativity theory. Thus, as it gains energy a particle gradually slips in phase relative to the fixed frequency accelerating electric field, and finally falls into the negative phase and is decelerated. In other words, above a certain maximum energy, the particle is no longer in resonance with the accelerating field.

Several tricks were invented to alleviate this effect, including the brute force method of increasing the voltage of the accelerating field to very high values so that a particle would carry out fewer revolutions in the machine, hence remain in resonance to higher energies. This limitation of fixed-frequency cyclotrons finally proved their downfall as the way to very high energies, but not before many heroic attempts were made to build larger machines and to develop better technology. The most remarkable attempt was made by Lawrence himself with the 184 inch cyclotron which was started just before the war at Berkeley, but was never completed as a fixed-frequency cyclotron. Fortunately, a new principle was discovered in 1945 which enabled this 184 inch cyclotron to be modified into a new type of cyclotron: for several years thereafter, it assumed a leading role as the highest energy particle accelerator in the world.

SYNCHROCYCLOTRONS

The new principle which opened the way to very high energy particle acceleration was announced almost simultaneously in 1945 by E. M. McMillan at Berkeley and V. I. Veksler in the USSR: it is called the principle of phase stability.

If the accelerating electric field could be varied in frequency to keep it in step with the particle rotation frequency, resonance could be extended to almost infinite energies, at least for a few particles, which consequently would be continuously accelerated. This might overcome the limitation of fixed-frequency cyclotrons.

The important question was whether a substantial number of particles would be held together as a group during acceleration, or whether the initial bunch of particles would disperse long before it reached high energies. The principle of phase stability showed that if a group or bunch of particles is centred around a certain phase of the accelerating voltage waveform, all the particles in the bunch oscillate stably about this equilibrium phase, and the whole bunch remains together up to the maximum energy that can be contained in the magnet of the machine.

The modifications required to a fixed-frequency cyclotron to make it into this new type of cyclotron, called a synchrocyclotron, are only to the accelerating system. Instead of a fixed-frequency system, a variable-frequency system is required. Usually a rotating capacitor is incorporated in the high-frequency circuit to vary its frequency.

Whereas the particle output of the fixed-frequency cyclotrons is continuous in time, and modulated only at the frequency of the accelerating electric field (i.e. about 20 MHz), the particle output of synchrocyclotrons is pulsed and bunches of particles emerge, again modulated at the frequency of the accelerating field, at a rate of about 100 bunches per second, determined by the rotation frequency of the capacitor.

The operation is briefly as follows. At each rotation of the capacitor a bunch of particles is captured from the source into a phase-stable 'bucket', and the bunch is then accelerated to the maximum energy of the machine. Finally, the particles are ejected from the machine onto external targets, or allowed to hit targets mounted inside the machine.

The particles during acceleration perform three motions at very different frequencies. First, they slowly spiral out as they gain energy from the particle source at the centre of the machine to the maximum radius following the equilibrium orbit. Second, they oscillate rapidly about the equilibrium orbit due to the focusing forces provided by the radial decrease in the magnetic field at the 'free', or betatron oscillation, frequency. Third, they oscillate slowly about the equilibrium orbit radially, and about the equilibrium phase angle azimuthally at the phase oscillation or synchrotron frequency.

The largest synchrocyclotrons built and operating are the modified 184 inch cyclotron at Berkeley giving a proton energy of 720 MeV, the machine at Dubna, near Moscow, giving 680 MeV protons, and the machine at CERN, giving 600 MeV protons. Larger machines could be built and the limitations were not so much in principle, as in such practical considerations as the high magnet cost and its large electrical power consumption.

The next step in the evolution of high-energy accelerators clearly had to avoid these excessive costs and power consumptions, and the way was found through the development of a new type of accelerator—the synchrotron.

SYNCHROTRONS

Synchrotrons differ from cyclotrons in that particle acceleration takes place at constant radius, rather than following a spiral path. Since the particle momentum is proportional to the product of the magnetic field and the particle orbit radius, acceleration at constant radius is possible only in a magnetic field which increases with particle energy, therefore with time. Because the particles circulate at constant radius the magnet can now take an annular form, hence, for a given maximum particle energy, it is less massive and much cheaper than the equivalent cyclotron magnet. Thus, the synchrotron concept allowed machines to be built for particle energies very much higher than was possible economically using synchrocyclotrons, and for the first time particle energies in excess of 1000 MeV or 1 GeV could be envisaged.

The requirement of a magnetic field increasing with time involved a new development in accelerators, namely the need to track the accelerating electric field frequency with the increasing magnetic field so as to preserve during the whole accelerating period a constant radius equilibrium orbit for the particles. Due to practical difficulties in making the magnetic field uniform around the equilibrium orbit at low magnetic fields, it was found necessary to start with a field of a few 100 gauss. Consequently, particles have to be injected into a synchrotron at an energy corresponding to the equilibrium radius of the machine and its initial magnetic field: for this purpose, another accelerator, called an injector, is required.

For the larger proton synchrotrons, giving output energies in the range of 3 to 10 GeV, the injectors used are either electrostatic machines or proton linear accelerators with output energies in the range 4 to 15 MeV. For the first time, therefore, accelerators were joined in series, and the particles, starting off in an ion source, were handed on from one accelerator to another until they reached their final energy.

Because the magnetic field in a synchrotron must increase with time from a few 100 G to 14 or 16 kG over a period of a few seconds, the magnet core can no longer be made of solid blocks of steel, as in the cyclotrons, but must be constructed of steel laminations glued or welded together to form magnet units, which are then arranged around the machine to provide an annular field.

The accelerating electric field in synchrotrons is usually provided by tunable resonant cavities, also distributed around the equilibrium orbit. The vacuum chamber enclosing the equilibrium orbit and mounted in the magnet gap, is also annular and the vacuum pumps, usually oil diffusion pumps, are distributed around the circumference of the machine.

The principle of phase stability used in synchrocyclotrons is equally necessary in synchrotrons. Also, the particle bunches are held together in the vertical and horizontal planes, as they circulate round the machine, by focusing forces provided by the curvature of the magnetic field lines. The particles during their acceleration in a synchrotron, therefore, circulate round the machine at constant equilibrium orbit radius, and oscillate about this equilibrium orbit in phase and energy in synchrotron oscillations, and horizontally and vertically in betatron oscillations.

Among the largest machines of this type built for accelerating protons are the 3 GeV Cosmotron at Brookhaven National Laboratory, completed in 1952, the 6 GeV Bevatron at Berkeley, completed in 1954, and the 10 GeV Synchrophasotron at Dubna, completed in 1957. In Western Europe, the largest proton synchrotrons are the 2·5 GeV Saturne machine at Saclay, completed in 1958, and the 7 GeV Nimrod machine at the Rutherford Laboratory near Oxford, completed in 1962.

Higher energy synchrotrons than these were designed during the early 1950s, but economic problems arose due to the weight and cost of even annular magnets, and the way to higher energies seemed once again blocked.

In 1952, however, a new principle was announced by Courant, Livingston and Snyder at Brookhaven for focusing the particles as they circulate round a synchrotron. This offered a way of reducing their amplitudes of oscillation about the equilibrium orbit, so that once again the cross-section dimensions of the magnet, hence its cost and power consumption, could be reduced.

ALTERNATING-GRADIENT SYNCHROTRONS

In both cyclotrons and synchrotrons, the focusing of the accelerating beam of particles around the equilibrium orbit is achieved by shaping the magnetic field lines so that the magnetic field all round the machine has a negative field gradient. Synchrotrons with this kind of focusing are often referred to as constant-gradient synchrotrons.

The new idea for focusing was to use a magnetic field gradient which is alternatively positive and negative round the circumference of the machine. Large synchrotron magnets are usually composed of several magnet units placed round the circumference of the machine to form the required annular magnet.

Appendix I/Particle Accelerators

According to the alternating gradient idea, there would be hundreds of such magnet units whose gradient would be positive and negative in successive units. Thus, a beam of particles circulating round such a machine would be focused and defocused in alternate units, and theory showed that the overall effect on the particles could be arranged to be focusing. Indeed, the focusing forces work out to be very much stronger with alternating-gradient magnets than with constant-gradient magnets, consequently the amplitudes of both the betatron oscillations and radial amplitudes of the synchrotron oscillations turned out to be very much smaller. Because of the strength of the focusing forces, the system is also referred to as the 'strong focusing' system.

The first proton synchrotron to be completed using this new focusing principle was the 28 GeV machine at CERN, which began to operate in 1959. It was closely followed by the 33 GeV proton synchrotron at Brookhaven, which came into operation in 1960.

Some idea of the economic advantage of using alternating-gradient focusing can be seen from comparing the weight of the magnets of the two types of synchrotrons. The steel core of the Bevatron magnets weighs roughly 10 000 tons, and that of the Russian synchrophasotron, 35 000 tons. Both these are constant-gradient machines.

On the other hand, the magnet core of the CERN machine weighs 3400 tons, and that of the Brookhaven machine 4000 tons. Since the CERN and Brookhaven machines give much higher proton energies than the Berkeley and Dubna Machines, the comparison is most striking if units of magnet core weight per GeV are used. In these units, the CERN and Brookhaven synchrotrons work out at 120 tons/GeV as against the 1500 tons/GeV of the Bevatron, and the 3500 tons/GeV of the Russian machine. Thus, the alternating-gradient idea gave reduction factors of 10 to 30 tons/GeV of magnet steel over the constant-gradient machines, and a reduction of about 4 in total machine costs.

However, these dramatic economic advantages were offset by certain technical problems. For example, the relatively small radial amplitudes of energy oscillation in alternating-gradient synchrotron produced a new phenomenon, hitherto not encountered in other accelerators, namely a critical energy during the acceleration process, called the transition energy. The reason is as follows. The principle of phase stability requires that a particle with more or less energy than the equilibrium particle adjusts its phase relative to the equilibrium phase, such that if it has excess energy it picks up less energy per revolution, or if it lacks energy it picks up more energy per revolution. The direction in which a particle moves in phase depends on two competing factors.

If it has energy in excess of the equilibrium energy, its velocity is greater than that of the equilibrium particle, which causes it to gain in phase over the equilibrium particle. However, because of its excess energy, it follows an orbit of greater radius than the equilibrium orbit radius, takes longer to complete a revolution, therefore falls back in phase relative to the equilibrium particle.

These two factors—velocity and radius—act in opposite directions. At the beginning of acceleration in an alternating-gradient synchrotron, the velocity factor predominates, hence determines the stable phase angle of the bunch of particles. As the bunch of particles gains energy, the effect of the velocity factor decreases as the particle's velocity approaches that of light, and at the transition energy, the velocity and radius factors just cancel out. There is then no phase stability.

After transition energy, the radius factor predominates, and phase stability is restored, but with the stable phase angle on the other side of the accelerating field waveform. It is, therefore, necessary to introduce a phase jump of about 60° at transition energy, and the great unknown when the CERN 28 GeV machine was being designed and built was whether the bunch of particles would remain together during the passage through transition energy when phase stability was lost.

The problem is rather analogous to a man running along with a cup in his hand containing hundreds of small balls. At a certain moment, he must remove the cup quickly leaving the balls in the air, and then put it back again so that none of the balls is lost.

Fortunately, the problem was eventually solved, and no loss of particles occurred in practice, but these worries resulted in many sleepless nights for the accelerator builders. The solution to this problem was greatly helped by the use of a phase-lock system, invented by Ch. Schmeltzer, now at Heidelberg, who was then in charge of this work at CERN. In the phase-lock system, which, incidentally, was the first use of a cybernetic system in accelerators, the bunch of particles automatically determines its own phase angle.

Other problems raised by the alternating-gradient idea concerned the extreme accuracy required of the magnet field. The magnet is composed of many units. Small errors in the magnetic field of these units, and in their physical lengths and position, give rise to a set of resonances which throw the particles out of the machine. Also, errors in the magnetic field gradients of these units give rise to another set of resonances, which have equally disastrous effects. It was, therefore, found essential to impose extremely tight tolerances on the permissible errors in the magnet units, to build very stable foundations for the machine, and to position the units round the machine circumference with very high precision.

Nevertheless, despite the transition energy problem and the extreme accuracy required in the manufacture and installation of the component parts of the accelerator, the alternating-gradient synchrotron has proved the most flexible and most efficient type of machine for the acceleration of protons to very high energies, and this design has now replaced all others in this field. Consequently, when a new generation of machines of higher energy than the 30 GeV accelerators was required for research, they were based on the alternating-gradient principle.

One of the frustrations of accelerating a beam of particles to high energies and then letting it hit a target is that so much of the energy, in a sense, is wasted as far as nuclear reactions are concerned. A high energy proton from an accelerator striking a proton at rest in a target produces only a fraction of its energy in the centre of mass system. Most of the energy is wasted in projecting the reaction products—the secondary particles—along the direction of the incoming proton. For example, the proton–proton centre of mass energy using the 28 GeV proton beam and a target is about 7 GeV, and this energy only increases as the square root of the primary beam energy.

A more efficient system would be to bring together two beams of protons of equal energy from opposite directions, since then the centre of mass energy would be just the sum of the energy of the two beams. Whereas a 28 GeV proton striking a proton at rest in a target gives 7 GeV centre of mass energy, two protons each of 28 GeV energy colliding together from opposite directions give 56 GeV centre of mass energy. To obtain 56 GeV centre of mass energy from a proton beam striking a target would require an incident beam energy of 1500 GeV.

The difficulty is the very low reaction rates which result from the collision of two beams of particles, due to the relatively low density of particles in the beams. To overcome this, it is necessary to build up several amperes of protons in each beam, and even then the unwanted background reactions due to protons hitting gas atoms in a vacuum of say 10^{-7} torr, which is usual in accelerators, masks the proton–proton reactions. Until methods of accumulating dense beams of particles were developed, and vacuum pressures could be lowered to 10^{-10} or 10^{-11} torr, this idea lay dormant in the minds of the accelerator builders.

In 1971, a very large machine of this type, called the Intersecting Storage Rings, came into research use at CERN. This uses the 28 GeV proton synchrotron to accelerate the protons which are then fed into each ring. To build up several amperes of protons in each ring, it is necessary to store in each ring about a thousand output pulses from the 28 GeV machine, and this takes about an hour of running time of the 28 GeV machine. Once the required current is built up in each ring, the two proton beams can be brought together at several points round the circumference of the machine, and in these regions the high energy proton–proton reactions take place.

Storage rings have been added to the DESY machine at Hamburg, and electron storage ring devices are in operation in France and Italy, so that in Europe we have both proton and electron storage rings.

If storage rings are so effective for reaching the very highest nuclear reaction energies, is it necessary to go on building high energy accelerators? Yes. First, the reaction rates are many orders of magnitude less using intersecting beams than using a single beam on a target, even with beam intensities in the rings of many amperes. Second, only proton–proton interactions can effectively be studied since the intensity of secondary particles from these reactions is very small indeed. Third, if the aim is to get the highest cosmic ray energies, it is necessary to start with an accelerator of energy much higher than 28 GeV to feed the rings.

This leads us to the story, as told in this book, of the building of the CERN 400 GeV accelerator, the SPS.

Appendix II/List of Contracts above 1 million Sw. Fr.

Contractor   *Subject*

Cockerill   Steel sheets for magnets
Chantiers Modernes   Civil engineering, North area
Morfax   Magnet cores
S.A.C.   Laboratories, offices, buildings
Losinger   Underground construction
Alsthom   Coils for bending magnets
BBC Baden   Magnets and power supplies
G.T.M.   Underground constructions
Plessey   Quadrupole and dipole magnets
Beton & Monierbau   Civil engineering, North area
Siemens   Power amplifiers
Seli   Underground construction
Lintott   Coils for bending magnets
Outokumpu   Copper for coils
Guffanti   Auxiliary buildings
BBC Mannheim   Bending magnets
Leybold   Vacuum pumps
York   Cooling towers
Hazemeyer   Low voltage switchboards
Norsk Data   Computers
Capag   Installation piping and cabling
Nordon   Piping installation
Voest   Cores for magnets
G.E.C.   Power compensator
Spie   Electrical installations
Brentford   Power supplies
Magrini   18 kV switchgears

Contractor	Subject
Saunier Duval	Electrical installations
Trindel	Installation of cables
Thomson	Power cables
Robbins	Tunnelling machine
C.D.R.	Piping installation
Borer	Electronic modules
S.L.E.	Electrical installations
Kabel & Metal	Water-cooled cables
TITN	Message transfer system
Pan Electric	Electric installations
Metall. Dornach	Copper tubes
Fagerstat	Stainless steel tubes
Sepa Levage	Overhead cranes
BCV	Services and equipment, North area
Cables de Lyon	18 kV cables
SEGCEM	Coils for bending magnets
Printhal	Auxiliary building
Raichon	Heating, ventilation
VAT	Vacuum
Spiezelectra	Filters
Sodeteg	Vacuum tubes
Gebauer	Lifting equipment
Van Swaay	Air conditioning installation
Hawker Siddeley	380 kV transformers
Alfa-Laval	Heat exchangers
CVM	Installation piping and cables
LTT	Electronics for beam monitors
Danfysik	Correction dipole magnets
Alpin	Drilling machine
Socea	Pumping station equipment
Zurnieden	Extraction towers
Smith & Johnson	Flanges, clamps for vacuum
CEA	Superconducting magnets
Sogeme	Vacuum installations
Muller	Mechanical supports

Appendix II/List of Contracts above 1 million Sw. Fr.

Contractor	Subject
Pfeiffer	Vacuum equipment
IKO	Multicore cables
Ensival	Pumps for cooling station
Hoesch Handel	Magnetic collimators
Birkle Thomer	Painting works
Puk Werk	Cable trays
Asea	Capacitors
O.T.E.	18 kV transformers
Messerschmitt	Measuring heads
Luwa	Air conditioning equipment

Acknowledgments

The authors would like to thank the Directors-General of CERN and the Group Leaders of the SPS Project for their help in preparing this book. Even at times of stress, the Group Leaders took pains to try to explain the complexities of their work in terms which could be understood by the less initiated. In addition, they made many invaluable suggestions as to how the draft chapters could be improved.

The photographs presented in this book and not individually acknowledged were, for the most part, taken by Gérard Bertin, Gilbert Cachin and François Julliard, and processed by Doris Wutschert of the CERN photographic section. We should like to express our appreciation of the contribution of this team which is daily facing the problem of portraying, on the one hand, huge constructions and, on the other, minute details of intricate equipment.

Special thanks are due to Brian Southworth, the indefatigable editor of the *CERN Courier*. Our work would have been much more difficult if he and his journal had not been there to consult.

Index of Names

Adams, J. B., 26, 34, 36, 38, 56, 58, 59, 69, 72, 75, 133, 134, 207, 211, 212, 216, 218, 236
Allaby, J., 60
Amaldi, E., 21, 24, 25, 37
Auger, P., 21, 22

Bachy, G., 207
Bakker, C. J., 36
Balke, S. M. S., 36
Bannier, J. H., 22, 61
Bernardini, G., 21, 25
Billinge, R., 115, 121, 133, 134, 212, 220
Blackburne, N., 75, 81
Blackett, P. M. S., 21
Bloch, F., 26, 30
Bohr, N., 21, 22, 24, 36
Bothe, W., 240
Brianti, G., 74, 205, 219, 223, 224
Burnod, L., 211

Cockcroft, J., 21, 27, 36, 243, 244
Coulomb, C. A. de, 240
Courant, E. D., 248
Crowley-Milling, M., 185, 186, 207, 212, 219

Dahl, O., 241
Dautry, R., 22
De Broglie, L., 21
De Raad, B., 74, 155, 205, 212, 214, 216, 220
Dockheer, F., 72

Fanfani, A., 49
Florent, R., 80
Flowers, B., 47, 51, 53, 59, 60
Fox, J., 137, 138
Funke, G., 27, 53

Gervaise, J., 74, 93, 96, 111, 129, 219
Göbel, K. J., 74, 81, 85, 219
Goldhaber, M., 33
Gregory, B. P., 38, 48, 53, 56, 59

Hampton, G. H., 39
Haunschild, H. H., 52
Heisenberg, W., 21, 33, 45
Henry, Y., 81
Hine, M., 52
Horisberger, H., 74, 81, 84, 207, 219, 225

Jentschke, W. K., 60
Johnsen, K., 42
Jonas, F., 49

Klein, A., 74, 76, 77, 220
Kohlhörster, W., 240
Kowarski, L., 21, 25
Kummer, W., 49

Lawrence, E. O., 244
Lee, T. D., 33
Le Vasseur, N., 78
Lévy-Mandel, R., 74, 84, 95, 97, 138, 147, 205, 219, 225
Livingston, M. S., 248
Lockspeiser, B., 22, 26

McMillan, E. M., 33, 36, 245
Martin, J., 51, 53
Merrison, A. W., 55
Milman, B., 74, 80, 205
de Modzelewska, E., 61

Oppenheimer, J. R., 22, 36

Paul, W., 52
Perrin, F., 21, 36, 51
Persson, L., 75, 80
Pickavance, T. G., 60
Picot, A., 22, 24
Powell, C., 33, 240

Rabi, I., 22, 33
de Rose, F., 22, 36
Rutherford, E., 239

Scherrer, P., 21, 24
Schmeltzer, Ch., 34, 250
Schnell, W., 34, 168, 178
Shaw, E. N., 60
Snyder, H., 248
Stassinakis, G., 78
Stoltenburg, G., 48

Thatcher, M., 59, 60
Toussaint, M., 48
Tröndle, P., 77, 78

Van de Graaf, R. J., 241
Van der Meer, S., 74, 138, 144, 212, 214, 217, 219
Van Hove, L., 38
Van Offelen, J., 48
Veksler, V. I., 245

Walton, E., 243, 244
Weisskopf, V. K., 38, 48
Wideröe, R., 241
Wilkinson, D., 52
Willems, J., 22, 26
Wilson, E., 72, 80, 205, 216, 220
Wüster, H.-O., 73, 75, 219

Yang, C. N., 33

Zettler, C., 72, 74, 166, 168, 178, 212, 214, 217, 220
Zilverschoon, C., 42

Subject Index

Accelerating cavities, 168–174, 247
Accelerating field, 35, 167, 174
Accelerating system, 87, 167
Acceleration, 165–180
Accelerators, 11, 34, 239–252
Access shaft, 101–103
Alternating gradient, 248
Amplifiers, 172
Atoms, 7, 9

Batavia, USA, 54, 115, 133, 207
Beam dump, 162, 164, 208
Beam monitors, 119, 182, 193
  instability, 182
BEBC, 80, 226, 229, 230
Bending magnets, 35, 116, 118, 123, 125, 134, 143, 193
Bubble chamber, 13, 17, 43, 226, 229, 230

CERN/1050, 86
CERN convention, 25, 28, 29
  Council, 30, 218
Čerenkov counter, 12, 16
Charm, 18, 230
Closed orbit, 181, 194
Commissioning, 205
Computers, 80, 184–204
Conseil Européen pour la Recherche Nucléaire (CERN), 22, 25
Continuous transfer, 154
Contract, 100
Control consoles, 199–201, 209, 217

Control of power supplies, 193, 197
Control room, 201–203, 209
Control System Working Group, 181
Cooling towers, 150
Correction magnets, 119, 212
Costs, 38, 42, 54, 58, 62, 220, 221
Council for Scientific Policy (UK), 47
Cycle time, 63
Cyclotron, 244

Daresbury, UK, 47, 186
Data module, 198
Doberdo, Italy, 50, 55, 215
Drensteinfurt, Germany, 50, 54, 58
Drift tubes, 168, 170, 171

Ejection of protons, 153–164
Electric field, 12
Electromagnetism, 20
Electron, 9, 10, 12, 18
Electron volt (eV), 11
Electrostatic deflectors, 162
Elementary particles, 1, 9
European Committee for Future Accelerators (ECFA), 37, 43, 58, 77, 221, 225
European Organization for Nuclear Research (CERN), 24, 26
Executive Groups, 81
Experimental Areas Working Group, 221, 224
Experiments Committee, SPS, 221

Finance Committee, 30
Focant, Belgium, 54
Focusing magnets, 35, 116, 118, 175, 178, 194
  see also Quadrupole magnets

Gargamelle, 18, 226, 229, 230
Génissiat, power station, 139, 140, 142
GESSS Committee, 86
Gigaelectron volt (GeV), 12
Göpfritz, Austria, 54
Gravity, 20

Hadrons, 17, 20, 223, 226, 228, 230, 231, 233, 235
Heat dissipation, 150
High-voltage sources, 240
Hydrogen, 10, 45, 87, 226, 230

Injection of protons, 153–164, 193
Instability, of proton beam, 182
Interaction, of particles, 10
Intersecting storage rings *see* ISR
ISR (Intersecting storage rings), 39, 41, 42, 88, 93, 117, 125, 168, 211, 251

Karlsruhe, Germany, 86

Laboratories, 97, 98
Le Luc, France, 51, 54
Leptons, 17, 20
Linac (linear accelerator), 87, 184, 185, 247

Machine Committee, 56, 72
Magnet insulation, 115, 118, 132–134
Magnets, *see* Bending magnets, Correction magnets, Focusing magnets, Main ring magnets, Septum magnets
Magnetic field, 12
Main ring magnets, 135
MC 60, 63, 70
Mechanical Design Group, 81
Message transfer system, 187, 190, 192–194, 196
Missing magnet concept, 70
Mole (boring machine), 111
Molecules, 8
Motor-generator set, 137
Multiplexor, 195, 196
Mundford, UK, 50
Muon, 17, 224, 228, 231, 233, 235

Navigation underground, 100, 107, 111
Neutrino, 9, 18, 20, 224–229
Nodal, 195, 198
North Area, 231, 233
Nuclei, 9

Particle, elementary, 1, 9
  physics, 1
Phase stability, 245, 248
Photon, 9
Physics, particle, 1
  high energy, 1, 19
Positron, 230, 237
Power supplies, 135–147
  auxiliary, 145
  control, 144, 193, 197
  consumption, 146
Projectiles, 5, 9
Proton, 10, 12, 17, 19
  source, 87
  synchrotron *see* PS

Subject Index

PS (proton synchrotron), 36, 39, 63, 68, 88, 91, 117, 137, 153, 185, 193, 209, 214

Quadrupole magnets, 118, 120, 121, 127, 129, 144, 153, 193
Quanta, 4
Quark, 18
Q-value, 116, 120, 154, 194, 214–218, 220

Radiation, 81, 85, 114, 202, 228
Radiation Group, 81
Reactive power load, 136
Rectifier stations, 143, 146
R.f. field, 153, 170
  power supplies, 179
Robbins boring machine, 100, 106, 112
Running-In Committee, 205
Rutherford Laboratory, UK, 47, 86, 115, 137, 188, 248

Saclay, France, 86, 248
Scientific Policy Committee, 30, 55
Science Research Council (UK), 47, 53, 59
Scintillation, 12
  counter, 12, 13, 225, 230
Septum, 153, 157–160
  magnet, 160, 162, 209
Serpukhov, USSR, 37, 40
Shafts, access, 101–103
  service (PGC), 103
Site Installations Group, 97

Site selection, 43
Spark chamber, 12, 15
Strong focusing, 249
Strong-focusing proton synchrotron, 34, 35, 63
Strong force, 20
Superconductive magnets, 65–68, 86, 232
  coils, 229
Synchrocyclotron, 32, 246
Synchrotron, 247
Target, 5, 10, 12, 226, 228, 233
Timing signal, 193
Touch screens, 201
Transition energy, 178, 250
Travelling wave accelerator, 167, 168
Tunnel boring, 100–114
  reinforcement, 108–110

UNESCO, 21, 22
United States, 22, 54
USSR, 22, 37

Vacuum chambers, 125, 127
Vengeron, 148

Water supplies, 148–152
  temperature, 151
Wave, 8, 10
Weak force, 20, 226
West Area, 225
  Hall, 230–232
Wimshurst machine, 241